André Filby

Interaction of colloids with mineral surfaces

André Filby

Interaction of colloids with mineral surfaces

a microscopical and nanoscopical approach

Südwestdeutscher Verlag für Hochschulschriften

Impressum/Imprint (nur für Deutschland/ only for Germany)
Bibliografische Information der Deutschen Nationalbibliothek: Die Deutsche Nationalbibliothek verzeichnet diese Publikation in der Deutschen Nationalbibliografie; detaillierte bibliografische Daten sind im Internet über http://dnb.d-nb.de abrufbar.
Alle in diesem Buch genannten Marken und Produktnamen unterliegen warenzeichen-, marken- oder patentrechtlichem Schutz bzw. sind Warenzeichen oder eingetragene Warenzeichen der jeweiligen Inhaber. Die Wiedergabe von Marken, Produktnamen, Gebrauchsnamen, Handelsnamen, Warenbezeichnungen u.s.w. in diesem Werk berechtigt auch ohne besondere Kennzeichnung nicht zu der Annahme, dass solche Namen im Sinne der Warenzeichen- und Markenschutzgesetzgebung als frei zu betrachten wären und daher von jedermann benutzt werden dürften.

Verlag: Südwestdeutscher Verlag für Hochschulschriften Aktiengesellschaft & Co. KG
Dudweiler Landstr. 99, 66123 Saarbrücken, Deutschland
Telefon +49 681 37 20 271-1, Telefax +49 681 37 20 271-0
Email: info@svh-verlag.de
Zugl.: Karlsruhe, Universität Karlsruhe, Dissertation, 2009

Herstellung in Deutschland:
Schaltungsdienst Lange o.H.G., Berlin
Books on Demand GmbH, Norderstedt
Reha GmbH, Saarbrücken
Amazon Distribution GmbH, Leipzig
ISBN: 978-3-8381-1370-8

Imprint (only for USA, GB)
Bibliographic information published by the Deutsche Nationalbibliothek: The Deutsche Nationalbibliothek lists this publication in the Deutsche Nationalbibliografie; detailed bibliographic data are available in the Internet at http://dnb.d-nb.de.
Any brand names and product names mentioned in this book are subject to trademark, brand or patent protection and are trademarks or registered trademarks of their respective holders. The use of brand names, product names, common names, trade names, product descriptions etc. even without a particular marking in this works is in no way to be construed to mean that such names may be regarded as unrestricted in respect of trademark and brand protection legislation and could thus be used by anyone.

Publisher: Südwestdeutscher Verlag für Hochschulschriften Aktiengesellschaft & Co. KG
Dudweiler Landstr. 99, 66123 Saarbrücken, Germany
Phone +49 681 37 20 271-1, Fax +49 681 37 20 271-0
Email: info@svh-verlag.de

Printed in the U.S.A.
Printed in the U.K. by (see last page)
ISBN: 978-3-8381-1370-8

Copyright © 2010 by the author and Südwestdeutscher Verlag für Hochschulschriften Aktiengesellschaft & Co. KG and licensors
All rights reserved. Saarbrücken 2010

Interaction of colloids with mineral surfaces:
a microscopical and nanoscopical approach

Zur Erlangung des akademischen Grades eines

DOKTORS DER NATURWISSENSCHAFTEN

von der Fakultät für

Bauingenieur-, Geo- und Umweltwissenschaften

der Universität Fridericiana zu Karlsruhe (TH)

genehmigte

DISSERTATION

von

Dipl.-Geol. André Filby

aus Karlsruhe

Tag der mündlichen
Prüfung: 11.11.2009

Referent: Prof. Dr. Bosbach

Korreferent: Prof. Dr. Fanghänel

Karlsruhe, 2009

Abstract

Deep geological formations are under serious consideration for the long-term storage and disposal of nuclear waste. Storage in crystalline rock in combination with a multi-barrier system is currently favoured in various countries as a feasible repository concept for high-level radioactive waste. However, in the case of possible contact with groundwater the release of radionuclides can not be excluded. These radionuclides might migrate into the environment by groundwater transport. Thus, the question arises as to what extent the radionuclides can reach the biosphere and which retardation mechanisms exist. The mobility of radionuclides strongly depends on sorption and desorption processes on the mineral surfaces of the surrounding rocks of the containment. Furthermore, the radionuclides may be adsorbed onto colloids possibly present in the groundwater and can thus be transported over considerable distances. For the long-term safety analysis of a nuclear waste disposal site it is, therefore, essential to investigate the interaction of colloids with mineral surfaces and to elucidate the processes responsible for the colloid-mineral surface interaction.

In field experiments carried out at the Grimsel Test Site (GTS), the migration behaviour of actinides in the presence and absence of bentonite colloids in a shear-zone was studied. Colloid recovery was in the range of 50-100 %. A high colloid recovery was expected, because in the Grimsel system generally repulsive conditions are apparent due to the high pH of Grimsel groundwater (pH 9.6). In the related laboratory experiments, the colloid recoveries were always lower than those found in the field studies (depending on flow velocity and colloid size the colloid recoveries ranged from 10 to 50 %).

Previous experiments have shown that the interaction of colloids as nanoscopic radionuclide carriers with natural mineral surfaces is complex. Comprehension of the processes involved is made difficult by, e.g., the chemical heterogeneity of the surfaces, surface roughness, discrete surface charges (on mineral edges or planes) or by surface contamination. Thus, the aim of this work was to gain insight into the interaction of negatively charged model colloids with natural mineral and rock surfaces. This was accomplished using different surface sensitive methods: sorption experiments were carried out with fluorescing carboxylated latex model colloids and Grimsel granodiorite and its main component minerals (quartz, biotite, muscovite, feldspar, apatite, titanite and additionally sapphire which was used as model for clay minerals and analogous iron phases). Sorption of the colloids was made visible by fluorescence microscopy. By using Scanning Electron Microscope (SEM) and energy-dispersive X-ray analysis (EDX) those mineral phases were identified on which predominant colloid adsorption took place. The latex colloids were chosen as model for natural bentonite colloids since both are negatively charged over a wide pH range and the difference of their surface potentials is not too high. However, one has to make a distinction regarding the comparability between natural and synthetic colloids: the spherical latex colloids do not exhibit the heterogeneous charge distribution typical of a natural clay colloid with its differing charges on edges and planes. Furthermore, it has to be noted, that colloid attachment and agglomeration are kinetically controlled phenomena. The short-term experiments carried out in this work are thus not transferable onto long time scales without further confirmation.

In the colloid adsorption experiments attractive interaction in the alkaline regime could only be observed on apatite. On all other mineral phases no colloid adsorption could be detected in alkaline conditions.

In an additional colloid desorption experiment with Grimsel granodiorite the colloids were initially adsorbed onto the rock surface at acidic pH. The desorption of the colloids was observed in alkaline conditions. The measurements showed that the colloid adsorption is reversible in alkaline conditions.

Complementary Atomic Force Microscopy (AFM) force spectroscopy experiments were undertaken with the colloid probe technique to gain insight into colloid-mineral surface binding mechanisms. With this method it was possible to identify and quantify attractive and repulsive forces resulting from the colloid-mineral surface interaction with high sensitivity. Both experimental approaches were conducted in a wide pH range and showed strong adsorption or attractive forces, respectively, at pH values close to or below the points of zero charge (pH_{pzc}) of the mineral surfaces. The influence of metal cations such as Eu(III) (used as chemical homologue for trivalent actinides), UO_2^{2+} (U is a main component of spent nuclear fuel rods and occurs under environmental conditions as mobile UO_2^{2+} ions), Ca(II) (as main component divalent cation in Grimsel groundwater) and natural Grimsel groundwater on colloid-mineral interaction was investigated. Depending on mineral phase and pH, a significant increase of colloid adsorption and attractive forces was observed in the presence of the cations due to their adsorption on the mineral and colloid surfaces. The experiments indicated that the adsorbed cations may reduce the overall repulsive interaction at pH values higher than the individual mineral pH_{pzc}, but still no colloid adsorption on most of the minerals (quartz, biotite, muscovite, feldspar, titanite, sapphire) was observed in the alkaline regime. In correspondence to the sorption experiments an attractive interaction in alkaline conditions could only be observed in the experiments with apatite and Grimsel groundwater or 10^{-4} M Ca(II).

Adhesion forces were significantly enhanced (and the magnitude of repulsive forces decreased) also at high pH in presence of 10^{-4} M Ca(II) ions and Grimsel groundwater.

The shear forces due to groundwater flow were estimated and found to be not high enough to overcome the measured colloid adhesion under all observed conditions. Detachment of adsorbed colloids due to ground water flow can thus be assumed as negligible.

To support the experimental results, the AFM measurements were compared to theoretical DLVO calculations. These calculations showed that the experiments can be well predicted by theory but that the calculations do not regard possibly present hydration forces which may prevent a possible colloid adsorption.

Results of this study prove that the interaction of colloids with mineral surfaces is determined by electrostatic interactions. According to the results, colloid adsorption in alkaline regime can only be expected on apatite in presence of Ca(II) or with Grimsel groundwater.

Zusammenfassung

Für die sichere Entsorgung nuklearer Abfälle sind tiefe geologische Formationen vorgesehen. Als mögliches Entsorgungskonzept für nukleare Abfälle wird die Endlagerung im Kristallingestein in Kombination mit einem Multibarrieren-System erachtet. Im Falle eines Wasserzutritts in ein nukleares Endlager sind die Freisetzung von Radionukliden und ihre Migration mit dem Grundwasser in die Umgebung wahrscheinlich. In diesem Zusammenhang ist es wichtig zu wissen, in welchem Umfang freigesetzte radioaktive Stoffe in die Biosphäre gelangen können und welche Retardierungsmechanismen existieren. Die Mobilität der Radionuklide hängt stark von Adsorptions- und Desorptionsprozessen an den Mineraloberflächen ab. Weiterhin können Radionuklide an natürliche Kolloide, die im Grundwasser ubiquitär vorhanden sind, sorbieren und so über beträchtliche Entfernungen transportiert werden. Für den Nachweis der Langzeitsicherheit eines nuklearen Endlagers ist es daher notwendig, die für die Kolloidwechselwirkung mit Mineraloberflächen zugrundeliegenden Prozesse aufzuklären. Im Felslabor Grimsel (Schweiz) wurden Feldversuche in einer wasserleitenden Scherzone durchgeführt, um das Migrationsverhalten von Aktiniden in An- und Abwesenheit von Bentonit-Kolloiden zu untersuchen. Hierbei lag der Wiedererhalt der Kolloide bei 50-100 %. Aufgrund der allgemein repulsiven Bedingungen im Grimselsystem (der pH-Wert des Grimsel-Grundwassers liegt bei 9.6) war der hohe Kolloidwiedererhalt ein erwartetes Ergebnis. In ähnlich gelagerten Laborversuchen wurden allerdings unerwartet niedrige Kolloidwiedererhalte beobachtet: in Abhängigkeit von Fließgeschwindigkeit und Kolloidgröße wurden Wiedererhalte von 10 bis 50 % gefunden. Diese Experimente zeigten auf, dass die Wechselwirkung von Kolloiden als nanoskopische Radionuklidträger mit Mineraloberflächen komplex ist. Ein Verständnis der zugrundeliegenden Prozesse wird durch die chemische Heterogenität und Rauhigkeit der Oberflächen, durch diskrete Oberflächenladungen (Ladungen an Flächen und Kanten) und durch Einflüsse von Oberflächenkontamination erschwert.

Ziel dieser Arbeit war es, die Wechselwirkung von Kolloiden mit Mineraloberflächen besser zu verstehen. Vor diesem Hintergrund wurden verschiedene experimentelle Ansätze verwendet. Im ersten Schritt wurden Sorptionsversuche mit fluoreszierenden carboxylierten Polystyrolkolloiden und Grimsel Granodiorit bzw. seinen Mineralbestandteilen (Quarz, Biotit, Muskovit, Feldspat, Apatit, Titanit und zusätzlich Saphir als Modell für Tonminerale und Eisenphasen) durchgeführt. Die Sorption der Kolloide wurde mittels Fluoreszenzmikroskopie sichtbar gemacht. Mit Hilfe der Rasterelektronenmikroskopie (REM) bzw. der energie-dispersiven Röntgenfluoreszenzanalyse (EDX) wurden jene Mineralphasen identifiziert, auf denen eine vorwiegende Kolloidadsorption stattfand. Carboxylierte Polystyrolkolloide wurden als Modell für natürlich vorkommende Bentonitkolloide gewählt, da beide über einen weiten pH-Bereich negativ geladen sind und sich ihre Zeta-Potentiale nicht zu sehr voneinander unterscheiden. Eine Einschränkung hinsichtlich der Übertragbarkeit ist dennoch gegeben: die sphärischen Polystyrolkolloide haben eine homogene Ladungsverteilung im Gegensatz zu Tonmineralen, bei denen die Ladung an Kanten und Basalflächen unterschiedlich sein kann. Weiterhin ist anzumerken, dass die Adsorption und Agglomeration von Kolloiden kinetisch kontrollierte Phänomene sind. Die im Rahmen dieser Arbeit durchgeführten Kurzzeitexperimente können daher nicht generell auf längere Zeitskalen übertragen werden.

Bei den Sorptionsversuchen konnte im alkalischen pH-Bereich lediglich auf Apatit (mit Grimsel Grundwasser) eine Kolloidadsorption festgestellt werden.

Bei einem zusätzlichen Kolloid-Desorptionsexperiment mit Grimsel Granodiorit wurden die Kolloide in einem ersten Schritt bei niedrigem pH auf die Gesteinsoberfläche sorbiert. Die Desorption der Kolloide wurde unter alkalischen Bedingungen beobachtet. Die Messungen haben gezeigt, dass die Kolloidadsorption unter alkalischen Bedingungen reversibel ist.

Um ein genaueres Verständnis über die Wechselwirkung von Kolloiden mit Mineraloberflächen zu erlangen und die für die Kolloidwechselwirkung verantwortlichen Kräfte zu messen, wurden Kraft-Abstands-Kurven mit dem Rasterkraftmikroskop aufgenommen. Dabei wurde die sogenannte colloid probe Technik verwendet. Mit dieser Methode konnten die für Wechselwirkung verantwortlichen Kräfte zwischen einem einzelnen carboxylierten Polystyrolkolloid und einer Mineraloberfläche mit hoher Empfindlichkeit identifiziert und quantifiziert werden. Beide experimentellen Ansätze wurden über einen weiten pH-Bereich durchgeführt und zeigten eine starke Adsorption der Kolloide bzw. starke attraktive Kräfte bei pH-Bereichen nahe oder unterhalb der Ladungsnullpunkte der jeweiligen Mineraloberflächen. Weiterhin wurde der Einfluss von Metall-Kationen wie Eu(III) (als chemisches

Homolog für dreiwertige Aktiniden), UO_2^{2+} (Uran ist eine Hauptkomponente von abgebrannten Kernbrennstäben und liegt im Grundwasser bei niedrigen pH-Werten als mobiles UO_2^{2+} vor), Ca(II) (als hauptsächlich vorkommendes Kation im Grimsel-Grundwasser) und Grimsel-Grundwasser auf die Wechselwirkung der Kolloide mit den Mineraloberflächen untersucht. In Anwesenheit dieser Kationen wurde - in Abhängigkeit von der Mineralphase und des pH - ein deutlicher Anstieg der Kolloidsorption bzw. der attraktiven Kräfte beobachtet. Die Experimente wiesen darauf hin, dass die adsorbierten Kationen (vor allem bei pH-Werten oberhalb der individuellen Mineral-Ladungsnullpunkte) die repulsive Wechselwirkung zwischen den wechselwirkenden Oberflächen erniedrigen. Dennoch konnte im alkalischen pH-Bereich auf dem Großteil der Minerale (Quartz, Biotit, Muskovit, Feldspat, Titanit, Saphir) keine Sorption der Kolloide bzw. keine auf die Kolloidadsorption hinweisende attraktive Kraft beobachtet werden. Wie bei den Sorptionsexperimenten konnte jedoch unter alkalischen Bedingungen auf Apatit mit Grimsel Grundwasser bzw. in Anwesenheit von 10^{-4} M Ca(II) eine attraktive Wechselwirkung beobachtet werden.
In Anwesenheit von 10^{-4} M Ca(II) wurde, im Vergleich zu den Experimenten mit Hintergrundelektrolyt, eine Erhöhung der Adhäsionskräfte (und Erniedrigung der repulsiven Kräfte) festgestellt.
Die von der Grundwasserströmung erzeugten Scherkräfte konnten abgeschätzt werden und es konnte gezeigt werden, dass diese nicht stark genug sind um die Adhäsionskräfte der Kolloide zu überwinden. Durch die Scherkräfte der Grundwasserströmung werden somit bereits adsorbierte Kolloide nicht wieder freigesetzt.
Um die Ergebnisse der experimentellen Kraft-Abstands Messungen zu untermauern, wurden zusätzlich theoretische Kraft-Abstands-Kurven nach der DLVO-Theorie berechnet und mit den experimentellen Daten verglichen. Die Berechnungen zeigten, dass die Experimente gut durch die Modellrechnungen beschrieben werden können, aber dass die DLVO-Theorie die wahrscheinlich vorhandenen repulsiven Hydratationskräfte nicht berücksichtigt, welche eine mögliche Kolloidadsorption verhindern könnten.
Die Ergebnisse dieser Arbeit zeigten, dass die Wechselwirkung von negativ geladenen Kolloiden mit Mineraloberflächen durch elektrostatische Wechselwirkungen kontrolliert wird. Den Ergebnissen zufolge kann mit einer Kolloidsorption unter alkalischen Bedingungen, wie sie im Grimsel-System herrschen, nur auf Apatit in Anwesenheit von Ca(II) oder Grimsel-Grundwasser gerechnet werden.

Meinen Eltern in Dankbarkeit gewidmet

Erklärung

Ich erkläre hiermit, dass ich die vorgelegte Dissertation selbstständig verfasst und mich keiner anderen als der von mir ausdrücklich bezeichneten Hilfsmittel bedient habe.
Ich habe die Grundsätze der Universität Karlsruhe (TH) zur Sicherung guter wissenschaftlicher Praxis in ihrer aktuell gültigen Fassung beachtet.

Danksagung

Die vorliegende Arbeit wurde im Zeitraum von April 2006 bis April 2007 unter der Leitung von Prof. Dr. Thomas Fanghänel und im Zeitraum von Mai 2007 bis Oktober 2009 von Prof. Dr. Dirk Bosbach am Institut für Nukleare Entsorgung im Forschungszentrum Karlsruhe durchgeführt. Ich danke Herrn Fanghänel für die Vergabe und Herrn Bosbach für die wissenschaftliche Betreuung der Arbeit.

Mein besonderer Dank gilt Herrn Dr. Plaschke für seine ständige Diskussionsbereitschaft, die produktive Zusammenarbeit sowie für seine wissenschaftliche und moralische Unterstützung. Herzlichen Dank für die hervorragende Betreuung!

Besonderer Dank gilt Herrn Prof. Dr. Geckeis für seine wissenschaftlichen Anregungen und die wertvollen experimentellen Ratschläge.

Ich bedanke mich bei Herrn Dr. Johannes Lützenkirchen und Herrn Dr. Thorsten Schäfer für ihre wissenschaftlichen Anregungen und ihre Diskussionsbereitschaft.

Ich bedanke mich bei Herrn Dipl.-Ing. Richard Thelen (Institut für Mikrostrukturtechnik, FZK) und Herrn Dr. Carlos Ziebert (Institut für Materialforschung, FZK) für die Bereitstellung der Rasterkraftmikroskope.

Bei Herrn Dr. Alexander Welle (Institut für Biologische Grenzflächen, FZK) möchte ich mich für die Bereitstellung des Kontaktwinkelmessgeräts bedanken.

Bei Herrn Florian Huber möchte ich mich für die geochemischen Modellrechnungen bedanken.

Für die Herstellung von diversen Meßzellen sowie für andere technische Angelegenheiten gilt mein Dank dem Team der Werkstatt des Institutes für Nukleare Entsorgung.

Weiterhin möchte ich mich bei allen namentlich nicht erwähnten Kollegen des Instituts für Nukleare Entsorgung danken, die mir mit dem sehr guten Arbeitsklima die Arbeit erleichtert haben.

Ich bedanke mich bei meinem Vater Dr. Gordon Filby für das aufmerksame Korrekturlesen dieser Arbeit. Weiterhin bedanke ich mich bei meinen Eltern, dass sie alles dafür getan haben, mir diesen Weg zu ermöglichen.

Ich bedanke mich bei Frau Eva Schulze und Herrn Dr. Ahmed Abdelmonem für die moralische Unterstützung.

Contents

I Introduction 1

II Literature review 4

1. Aquatic colloids 4
    1.1 Interaction between colloids and metal cations 5
    1.2 Adsorption of metal ions onto inorganic colloids 5
    1.3 Colloid mediated transport of metal ions in subsurface environments 7
    1.4 Minerals and their surface properties 9

2. Colloid and surface forces 11
    2.1 The electrical double layer 11
    2.2 Van der Waals interactions 13
        2.2.1 Microscopic model according to Hamaker 14
    2.3 Aggregation and stability of colloidal dispersions 17
    2.4 Van der Waals and double-layer forces acting together: the DLVO theory 20
    2.5 Non-DLVO forces 22
    2.6 Hamaker constants and contact angle measurements 23
        2.6.1 Three-phase equilibrium and the Young-Dupré equation 24
        2.6.2 Interpretation of the contact angle measurements 25
        2.6.3 Interrelation between surface energy and Hamaker constant 26

III Methods and Materials 28

1 Methods 28
    1.1 Determination of zeta-potential 28
        1.1.1 Electrophoresis 28
        1.1.2 Streaming Potential 29
    1.2 Photon Correlation Spectroscopy (PCS) 30
    1.3 Contact Angle Measurements 32
    1.4 Fluorescence Microscopy 33
    1.5 Scanning Electron Microscopy (SEM) 35
    1.6 Atomic Force Microscopy (AFM) 36
        1.6.1 Cantilever and detection 37
        1.6.2 Measuring modi 38
        1.6.3 AFM force spectroscopy 40

2 Materials 45

2.1 Colloids 45
2.2 Sample solutions 46

3 Sorption and force spectroscopy experiments 48
3.1 Sorption experiments with Grimsel granodiorite 48
3.2 Single minerals used in sorption experiments 49
3.3 Single minerals used in force spectroscopy experiments 50

IV Results 51

1. Zeta-potential and PCS measurements 51
1.1 Fluorescent polystyrene colloids 51
1.2 Polystyrene spheres used as AFM colloid probes 53
1.3 Streaming potential measurements 54

2. Sorption experiments with fluorescent polystyrene colloids 55
2.1 Surface coverage measurements 55
2.2 Interaction of carboxylated latex colloids with component minerals of Grimsel granodiorite 56
    2.2.1 Sheet silicates (muscovite and biotite) 57
    2.2.2 Feldspars (albite and K-feldspar) 59
    2.2.3 Quartz 61
    2.2.4 Apatite 62
    2.2.5 Sapphire and titanite 64
2.3 Interaction of carboxylated latex colloids with Grimsel granodiorite 65
    2.3.1 Colloid desorption experiment 67

3. AFM force spectroscopy experiments 68
3.1 Sheet silicates (muscovite and biotite) 69
3.2 K-feldspar 74
3.3 Quartz 77
3.4 Apatite 79
3.5 Titanite 81
3.6 Force-volume measurements 82

4. Calculation of DLVO interaction force profiles 84
4.1 Contact angle measurements, surface energies and Hamaker constants 84
4.2 DLVO calculations 85
    4.2.1 Theoretical and experimental AFM force-distance curves 86
    4.2.2 Maximum potential energy barrier calculations 92

V Discussion 94
VI Summary and Conclusions 107
VII Literature 110
VIII Appendix...134
1. List of figures 135
2. List of tables...139
3. List of symbols 141
4. Acronyms 144

I Introduction

Generation of energy by nuclear fission leads to a production of spent fuel elements which contain uranium, plutonium, fission products and the so-called "minor actinides" such as neptunium, americium and curium. The safe disposal of this highly toxic and radioactive waste demands its isolation from the biosphere for several hundred thousand years. The radiotoxicity is a measure for the poisonous effect of radionuclides and is given in units of Sv/tHM (Sievert/ton of heavy metal). The radiotoxicity depends on the nature of the radiation, radiation energy, resorption of a radionuclide in the organism, biological effectiveness and its residence time in the human body. Fig. 1 shows the radiotoxicity as a function of time from ten to one million years. It can be seen that the radiotoxicity is dominated by the fission products in the first 300 years. Due to radioactive decay their radiotoxicity decreases and after ca. 300 years it is dominated by americium and plutonium; after 2000 years plutonium contributes mainly to the radiotoxicity. After 100000 years the daughter products from the decay of plutonium and the minor actinides provide the main contribution to radiotoxicity. Thus, the immobilization of long-lived radionuclides (actinides and fission products) over a geological timescale is the primary aim of nuclear waste disposal [1].

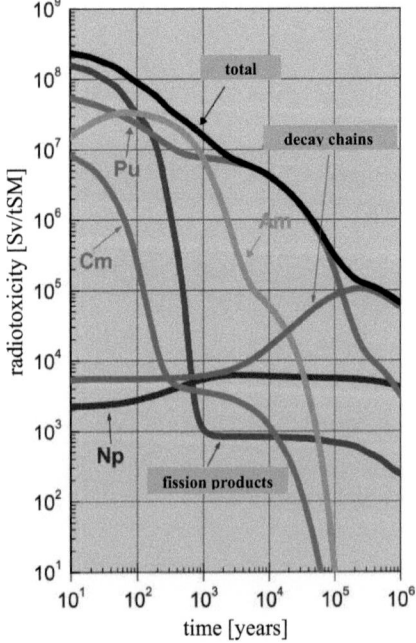

Fig. 1: Radiotoxicity of radionuclides in the spent fuel in dependence of time (see text)

Although many other concepts have been considered, storage in deep geological formations in combination with a multi-barrier system is currently considered as the most feasible repository concept for high-level radioactive waste [2,3]. The multi-barrier system is composed of the following [1]:

- the waste matrix itself consisting of the spent fuel and its container (technical barrier)

- the geotechnical barrier consisting of the backfill material such as compacted bentonite or salt, depending on the waste disposal concept

- the geological barrier (e.g., salt, granite, clay) and the aquifer in the overburden

For each of these barriers those processes have to be examined and described which lead to the mobilisation or immobilisation of radionuclides. A variety of processes can take place in the vicinity of the individual barriers (e.g., redox reactions, hydrolysis, sorption/desorption, colloid generation) and each has to be considered. The aim is to determine rates for the mobilisation and immobilisation of the radionuclides at the individual barriers and from this it is possible to derive the source terms for radionuclide release at the barriers. The source terms allow a quantification of the migration of long-lived radionuclides into the biosphere and hence it is possible to give a geochemically well-founded long-term safety analysis [1].

Bentonite clay has been found to be an appropriate material for the geotechnical barrier of the multi-barrier system due to its swelling properties and ability to adsorb radionuclides from solution. However, depending on the physico-chemical conditions (e.g., gas and water pressure, ground water flow velocity) the bentonite barrier may also generate colloids [4,5,9]. Colloids show, in comparison with ionic species, a different migration behaviour: due to their comparatively larger size and charge they can not intrude into pores of the host rock and hence may be transported unretarded or even faster than the average velocity of water which can intrude into pores. Thus, colloidal bound contaminants such as radionuclides may interact to a lesser degree with the surface of the host rock minerals and can be transported over considerable distances [6,7]. However, colloids can also be retained by interaction with mineral surfaces or by agglomeration, sedimentation and filtration. Both processes strongly depend on geochemical parameters, e.g., pH, ionic strength or colloid concentration. In the colloid and radionuclide retardation (CRR) experiment carried out at the Grimsel Test Site

(GTS) the in situ migration behaviour of selected actinides in the presence and absence of bentonite colloids in a water-conducting shear zone was studied [7,8]. Under fast flow conditions recovery of injected bentonite colloids was in the range of 50-100 %. In related laboratory experiments carboxylated polystyrene (latex) colloids, used as a model for bentonite colloids, were injected into a granodiorite bore core. Colloid recoveries were always lower than those found in the field studies. Depending on colloid size and flow velocity, colloid recoveries ranged from 10 to nearly 50 % [9]. Similarly low colloid recoveries with increasing residence time were found by other investigators [10]. The authors conclude that even under geochemical conditions where colloids are highly stabilized (low salinity, high pH) adsorption/filtration onto mineral surfaces may occur.

The extent of colloid adsorption on natural mineral surfaces may be influenced by, e.g., the chemical heterogeneity of the mineral surfaces, surface roughness, surface contaminations, discrete surface charges (mineral edges and planes), mineral dissolution, the presence of dissolved ions or by matrix diffusion [11,12,13,14,15,16]. Alonso et al. [17] studied the interactions between gold colloids and a granite surface by μ-Particle Induced X-Ray Emission to determine the amount of adsorbed colloids on different minerals under varying chemical conditions. They concluded that adsorption of colloids on rock surfaces is generally determined by electrostatic interactions. However, colloid adsorption was also detected under conditions under which the rock surface is unfavourable (repulsive) for a colloid attraction (e.g., alkaline pH). It was suggested that chemical effects may enhance colloid/rock interaction when favourable (attractive) electrostatic interaction does not exist. Furthermore, theoretical studies have been performed to calculate surface charge heterogeneity effects (e.g., [18]). These effects have also been experimentally evaluated, e.g. for Cu ion adsorption onto silica [19].

From the present status of knowledge, binding mechanisms of negatively charged colloids on natural mineral surfaces are not sufficiently understood. It can be expected that electrostatic interaction will dominate colloid adsorption. However, it is not clear if additional binding mechanisms are present, especially under unfavourable (repulsive) electrostatic conditions. This work intends to answer this open question.

In the preliminary phase of the study, adsorption experiments with carboxylated (negatively charged) fluorescent polystyrene colloids (takes as well characterised model particles) were performed on original Grimsel granodiorite fracture filling material and its single component minerals. The adsorbed colloids were easily identified on the sample surfaces by fluorescence microscopy and the mineral phases on which a predominant colloid adsorption took place were determined by SEM/EDX. In a further step, Atomic Force Microscopy (AFM) force spectroscopy measurements were carried out under varied geochemical conditions to measure interaction forces between model colloid and mineral surfaces. These measurements allow the identification and quantification of the individual forces responsible for the mineral-colloid interaction. Finally, the experimental results are compared to theoretical calculated force curves in order to determine mineral surface potentials and to find out if the experimental results can be predicted by theory.

II Literature review
1. Aquatic colloids

The word "colloid" is derived from the Greek word for glue ("κωλλα") and was introduced in 1861 by Thomas Graham (1805-1869), the first president of the Chemical Society of London [20]. For the range of the colloid dimension different definitions exist. Some authors classify colloids as particles with a diameter of 1 nm to 0.5 µm [20,21] or from 1 nm to 0.45 µm [22]. Sposito classifies colloids as solid particles with a diameter of 0.01 µm to 10 µm [23]. The current definition of the "International Union of Pure and Applied Chemistry" (IUPAC) establishes that dimensions of colloidal systems to be between 1 nm and 1 µm [24]. Fig. 2 shows the dimensions of the particles frequently found in aqueous systems. Inorganic groundwater colloids can be clay minerals, e.g. aluminium or magnesium layered silicates, iron- or aluminium oxides and other complex minerals [25]. The composition of these particles is dependent on the groundwater chemistry and the geomatrix [26]. The formation of such colloids can occur by the following processes: (1) through condensation from molecules or atoms (i.e. Al, Fe); (2) physical or chemical erosion of macroscopic forms (e.g., minerals or organic matter) [27,28]. In soil- and groundwater, there are also colloids with (mainly) organic composition. Besides microorganisms and viruses, natural organic matter (NOM) such as humic colloids consisting of humic substances can also be present [29,30].

A colloidal dispersion consists of grains or droplets of one phase in a matrix of another phase and is thus a two-phase system which is uniform on the macroscopic but not on the microscopic scale. The interface-to-volume ratio of colloids is so large, that their behaviour is mainly determined by surface properties. Inertia and gravity can be neglected in the majority of cases. Thus, colloidal systems are controlled by interfacial effects rather than bulk properties. Movement of colloidal particles is controlled by Brownian motion. This separates them from granular matter which is composed of macroscopic particles and is not subject to thermal motion [92].

II Literature review

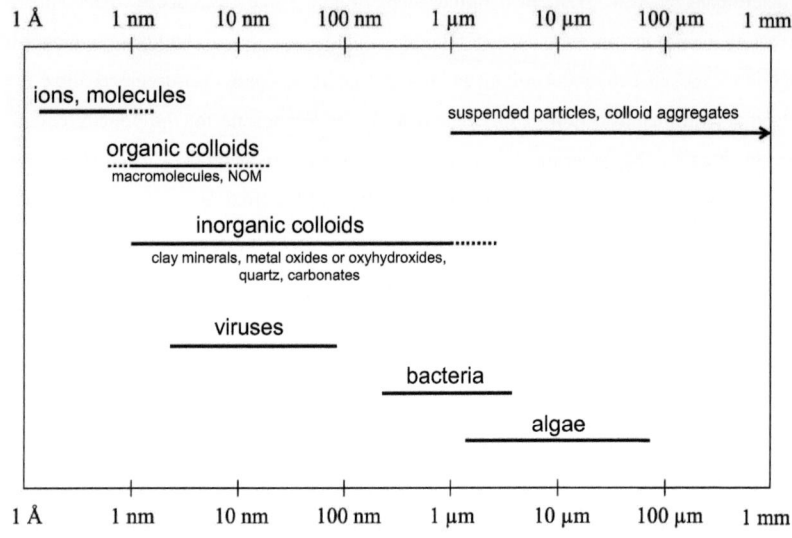

Fig. 2: Size distributions of particles common in natural aquatic systems (see text)

1.1 Interaction between colloids and metal cations

Colloid-metal cation interactions belong to the most important processes in subsurface environments having strong influence on the transport of the different cations present in soil- or groundwater. Metal cations can be adsorbed onto the colloid surfaces; they can replace other ions with similar charge (ion exchange) or can form complexes with organic/inorganic matter. These processes are dependent on several factors, such as the specific surface area of the colloidal particles, the density of the surface functional groups, the cation exchange capacity (CEC), the point of zero charge (pH_{pzc}), pH, ionic strength of the soil or ground water as well as the temperature. All these processes are reviewed in more detail in the literature, e.g., [31].

1.2 Adsorption of metal ions onto inorganic colloids

Regarding the mobility of metal ions in subsurface environments, the ubiquitously present inorganic colloids such as clay minerals, metal oxides or oxihydroxides and quartz play an important role. The sorption of ions onto their colloidal surfaces can result from van der Waals interactions or the formation of ionic or covalent chemical bonds. For the characterisation of the different sorption processes various surface complexation models were developed and applied, e.g.: Constant Capacitance Model [32,33,34], Basic Stern Model [35,36,37,38], Double Layer Model [39,40,41,42], Triple Layer Model [43,44,45,46], Charge Distribution Multi Site Complexation (CD-Music) Model [47,48,49,50]. These models differ mainly in the mathematical formulation of the double layer but in case of the CD-Music approach the arrangement of surface hydroxyl groups is derived from the mineral structure.

Ions can be adsorbed onto colloid surfaces as outer- or inner-sphere complexes [51,52] as depicted in the example of $\equiv$XOH surface functional groups in Fig. 3 (where $\equiv$X represents the lattice bound metal cation of the (hydr)oxide). Outer-sphere complexes (also known as unspecific adsorption) involve hydrated cations or anions which are bound as an ion pair by electrostatic attraction with the $\equiv$XO$^-$, $\equiv$XOH or $\equiv$XOH$_2{}^+$ surface functional groups (Fig. 3 (a)).

Regarding the adsorption as inner-sphere complexes (also known as specific adsorption) ligand exchange between the colloid surface and the adsorbing cations/anions takes place. Here, one (monodentate) or two (bidentate) ligands of a sorbing ion is bound on one (mononuclear) or two (binuclear) atoms of the colloid surface (Fig. 3 (b)-(d)). The formation of inner-sphere complexes proceeds with the release of H$^+$ (cation adsorption) or OH$^-$ (anion adsorption) from the surface.

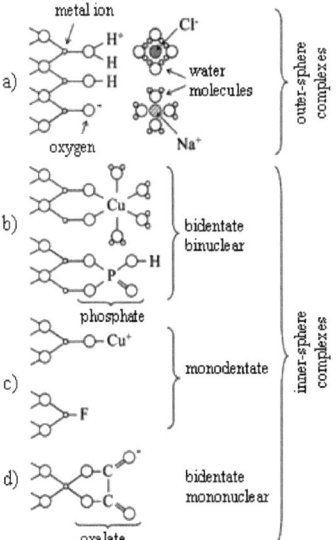

Fig. 3: Schematic illustration of the adsorption of cations/anions onto a colloid surface as inner- or outer-sphere complexes (modified after Scheffer & Schachtschabel [51]) (see text)

The adsorption of ions onto the colloid surface and the formation of surface complexes depends strongly on pH. At high and low pH values, the surface groups are deprotonated or protonated, respectively. Consequently the attractive electrostatic forces between the outer-spheric cation- or anion complexes and the negatively or positively charged surface functional groups are strong. For the inner-sphere complexes the pH also plays a significant role. The increase of pH promotes the formation of inner-sphere cation complexes (see equations 1 and 2), whereas a decreasing pH promotes the formation of inner-sphere anion complexes (equation 3) [52].

$$\equiv XOH + Cu^{2+} \leftrightarrow \equiv XOCu^+ + H^+ \tag{1}$$

$$2(\equiv XOH) + Cu^{2+} \leftrightarrow (\equiv XO)_2Cu + 2\,H^+ \tag{2}$$

$$2(\equiv XOH) + H_2PO_4^- \leftrightarrow (\equiv X)_2PO_4^- + H_2O \tag{3}$$

1.3 Colloid mediated transport of metal ions in subsurface environments

Past research regarding the subsurface transport of substances often disregarded the presence of colloids. Aquifers were considered as two-phase systems (solid phase: rock or soil matrix; fluid phase: subsurface water with diluted substances). More recent research has shown that colloidal particles play an important role in the transport of various substances in soil and groundwater aquifers [53,54,55,56,57,58]. Due to their large specific surface area and cation exchange capacity colloids can adsorb or bind contaminants such as heavy metals [55,56,57,59], radionuclides [55,58,60,61] and organic substances [55,57,60]. The colloid mediated transport is a complex process comprising various reaction steps: interactions between metal cations and colloids (ad- and desorption), precipitation and dilution processes of the metal ions and interaction of the colloids with each other (aggregation, adsorption of NOM onto the inorganic colloids) and with the surrounding rock or soil matrix (deposition or release). Fig. 4 shows a simplified scheme.

II Literature review 8

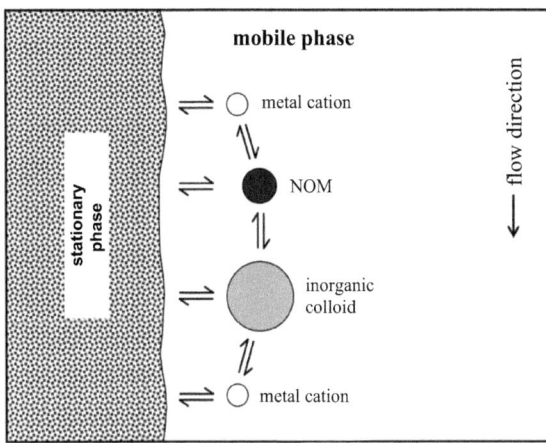

Fig. 4: Scheme of colloid-mediated transport and colloidal interaction in subsurface environments (see text)

All these interactions determining colloid-mediated transport or retention are influenced by various factors, i.e., pH, ionic strength of the fluid medium, surface charges, hydrophobicity of the inorganic and organic colloids and the hydraulic conditions. Inorganic colloids are generally very mobile at low ionic strength as a result of dominating repulsive forces between colloids and the stationary phase. Column-experiments showed that kaolinite colloids are very mobile at low ionic strength and that the transport of kaolinite sorbed cesium is strongly increased [58]. With increasing cation concentration in the aqueous phase the electrostatic attractive force between inorganic colloids and surrounding rock/soil matrix becomes stronger and the colloid deposition rate increases [55,62,63,64]. The adsorption of colloids onto the stationary phase is also dependent on the cation valency. The deposition rate of the colloids is significantly higher in systems with higher concentrations of two-valent cations (e.g., Ca(II)) than in systems where mono-valent cations (e.g., Na(I)) dominate [55,56]. In terms of the transport of colloids, the so-called "blocking" and "ripening" effects are important. When the repulsive forces between colloids are dominant, an increasing amount of the adsorbed particles leads to a decrease of the deposition rate of the colloids. This effect is commonly known as the "blocking" effect. In systems in which attractive forces between the colloidal particles dominate, a multi-layer adsorption of the colloids onto the surface of the stationary phase takes place. As a consequence, the deposition rate of colloids increases with increased number of adsorbed particles. This effect is known as "ripening" effect [55,62,65,66,67]. The colloid mobility is also strongly dependent of the pH. With increasing pH the surface of the colloidal particles and the stationary phase will become increasingly deprotonated. This leads to increasing repulsive forces between the corresponding negatively charged surfaces and the mobility of the colloids increases [54,68]. On the other hand, the sorption of, e.g., heavy metals onto the colloids increases with increasing pH [69,70,71,72].

The transport of colloids can also be influenced by NOM. Dissolved NOM, e.g., humic substances, can act as strongly complexing ligands. Humic substances appear in many groundwaters as organic degradation products of biological processes. In groundwater, humic substances may be present as polydisperse colloidal macromolecules with partly adsorbed inorganic constituents. Polyvalent heavy metal ions such as Zr(IV) and Th(IV) and the trivalent lanthanides are stabilized as humic colloids in natural groundwaters. Laser spectroscopic investigations proved that actinide ions form complexes with humic acids. It was shown that ternary complexes can be formed at alkaline pH in the presence of other aquatic ligands, such as carbonate- and hydroxide ions. Humic acids dominate the actinide speciation over a wide pH range because of the formation of such ternary complexes. Furthermore, the colloidal character of the aquatic humic substances causes -in comparison to the ionic species- a different migration behaviour. Colloids cannot intrude into the pores of the host rock due to size- and charge exclusion and can thus migrate unretarded or even faster than water itself. Colloidal bound radionuclides interact to a lesser degree with the surface of the geomatrix and thus the possibility of retardation may be decreased. Due to conformational changes in the humic acids as a result of the complexation with metal ions, significantly reduced actinide dissociation kinetics were observed. This indicates that over longer time periods formation of metaloxide/hydroxide clusters may take place. These may be stabilized by humic colloids and bind the actinide ions in an irreversible fashion [1].

1.4 Minerals and their surface properties

Oxides (especially those of Fe, Si, Al) are abundant components in the earth's crust. Thus, oxides and hydroxides are ubiquitous in natural waters, sediments, soils and rock. In presence of water, these oxides are covered with surface hydroxyl groups as reactive sites. The density of surface hydroxyl groups depends on the mineral structure. The hydroxyl-groups react amphoterically, e.g., as acid or base, depending on pH and the geochemical conditions. Fig. 5 (a) shows a schematic representation where metal ions in the surface layer have a reduced coordination number and thus behave as Lewis acids. In the presence of water the surface metal ions first tend to coordinate H_2O molecules. For most of the oxides the dissociation of

protons seems energetically favoured and OH-groups are formed which cover the surface (Fig. 5 (b)) [52].
A more detailed discussion about the individual mineral surface functional groups can also be found in section V.

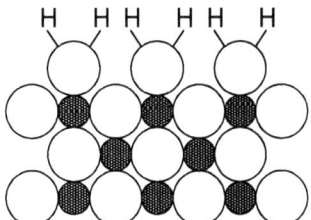 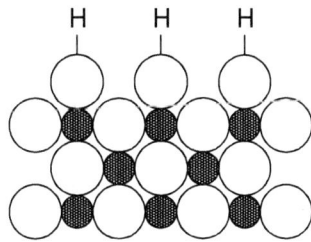

Fig. 5 (a) and (b): Schematic representation of the cross section of a metal oxide surface layer. Black spheres represent metal cations, white spheres are oxide ions [52] (see text)

The pH value at which the number of negatively charged functional groups equal the positively charged functional groups, is called the point of zero charge (pH_{pzc}). The pH_{pzc} of a mineral surface is routinely measured by acid-base titrations of powder suspensions or electrophoretic measurements [31]. On natural surfaces the charges can be distributed heterogeneously resulting in domains of high local charge density. These charge heterogeneities were observed on, e.g., quartz [19]. Also natural clay particles and sheet silicates are known to show surface charge heterogeneity. Due to the individual mineral structure and composition, the mineral edges can exhibit different pH-dependent charge as compared to the basal planes [73].

The Grimsel granodiorite used in this work was derived from fracture filling material taken from the GTS, Switzerland, where a wide range of research projects on the disposal of radioactive waste and radionuclide migration have been initiated (www.grimsel.com). The mean bulk composition of Grimsel granodiorite fracture filling material and the individual point of zero charges (pH_{pzc}) of the individual component minerals can be seen in Table 1 [74].

Table 1: Grimsel granodiorite bulk composition

Mineral	Vol.-%	pH_{pzc}
quartz, SiO_2	15	2^a - 3.8
plagioclase/albite, $CaAl_2Si_2O_8$/$NaAlSi_3O_8$	28	2^a, $5.25^{b,c}$
K-feldspar, $KAlSi_3O_8$	7	2-2.4^a, 5.6^b
biotite, $K(Mg,Fe)_3[(OH)_2/Si_3AlO_{10}]$	41	6.5^d
muscovite, $KAl_2[(OH)_2/Si_3AlO_{10}]$	4	4^d, 6.6^c
epidote, $Ca_2Al_2(Al, Fe^{3+})OOH[Si_2O_7][SiO_4]$	3	?
titanite, $CaTiOSiO_4$	2	?
chlorite, $(Mg,Fe,Al)_3(Si,Al)_4O_{10}(OH)_2(Mg,Fe,Al)_3(OH)_6$	< 1	?
apatite, $Ca_5[(F,Cl,OH)/(PO_4)_3]$	trace	7.6^e,8.1^f
rutile, TiO_2	trace	?
zircon, $ZrSiO_4$	trace	?
ilmenite, $FeTiO_3$	trace	?
orthite, $(Ca,Mn,Ce,La,Y,Th)_2Al(Al,Fe^{3+})(Fe^{2+},Fe^{3+},Ti)OOH[Si_2O_7][SiO_4]$	< 1	?
clay minerals	0-1	4-6^e

^{a} from [75], ^{b} from [76], ^{c} from [77], ^{d} from [78], ^{e} from [79], ^{f} from [80]

Note that the mineral composition of the surface can significantly differ from the bulk rock due to weathering of the minerals and possible deposition of alteration products on the surface.

2. Colloid and surface forces

Important properties of colloidal systems such as their stability and aggregation behaviour are determined directly or indirectly by the interaction forces between surfaces. Colloidal forces are composed of several components, such as electrical double layer-, van der Waals-, Born-, hydration- and steric forces, all of which depend on the surface properties of the particles or minerals [81]. If attractive forces dominate, particles aggregate, if repulsive forces dominate, the colloid dispersion is stable.

2.1 The electrical double layer

In water, most surfaces are electrically charged. Surface charges at the mineral-water interface can result from: (1) isomorphic substitution in the crystal structure of the individual mineral or (2) a pH-dependent protonation/deprotonation of surface functional groups (e.g. hydroxyl groups), known as variable charge. The isomorphic substitution of higher-valent by lower-valent cations within the crystal lattice forms a permanent negative charge [31]. Regarding the variable charge, acidic conditions lead to an excess of adsorbed H^+ which results in a net positive charge at the surface hydroxyl functional groups. In contrast, high pH conditions induce hydroxyl deprotonation with the surface gaining a net negative charge [82]. The surface ionization reactions responsible for the amphoteric behaviour of the surface hydroxyl groups can be described by the following expressions.

$$\equiv XOH_2^+ \leftrightarrow \equiv X(OH)^0 + H^+ \qquad K_{a1}^{app} \qquad (4)$$

$$\equiv X(OH)^0 \leftrightarrow \equiv XO^- + H^+ \qquad K_{a2}^{app} \qquad (5)$$

K_{a1}^{app} and K_{a2}^{app} are apparent equilibrium constants for reactions in equations 4 and 5, or apparent acidity constants for surface species $\equiv XOH_2^+$ and $\equiv X(OH)^0$ [83].

In an electrolyte solution, the distribution of ions around a charged particle is not uniform and gives rise to an electrical double layer. The particle surfaces are balanced out by an equivalent number of oppositely charged counterions in solution where they are subject to two opposing influences: electrostatic attraction tending to localize the counterions close to the particles and the tendency of ions to diffuse randomly throughout the solution in dependence of their thermal energy. The surface charge on a particle and the associated counterion charge together constitute the electrical double layer [81].

When two charged surfaces approach each other in an electrolyte solution and the electrical double layers overlap, an electrostatic double-layer force arises which can stabilize dispersions. Different factors influence the way in which the double layers interact. Distinction is made between interaction at constant surface potential or at constant surface charge [81]. The first case refers to the maintenance of surface-chemical equilibrium during approach. Due to the very short time of an collision between colloidal particles (about 10 μs for Brownian collision) this may not be a realistic assumption [84]. Constant-charge interaction can be expected when the particles have a fixed surface charge density, e.g., latex particles with ionic groups or clays with a defined ion exchange capacity. However, it should be noted, that neither of these extreme cases is likely to apply in practice [81].

The double layer interaction energy can be calculated in two ways. One is to solve the so-called Poisson-Boltzmann equation (for more details, see the literature: [81,91,95]), a partial differential equation of second order which in most cases has to be solved numerically [92]. The other method is to construct approximations from the known expressions for each of the surfaces involved in the absence of the others. For this work, the sphere-plate interaction is the most relevant. A simple expression, which represents a useful compromise between the two extremes of "constant potential" (mobile charges that keep the potential between the two surfaces constant) and "constant charge" (assuming immobile charges), is the so-called linear superposition approximation (LSA) which gives intermediate values between both cases. The condition of constant surface potential is considered to exist if the surface charge is determined by the adsorption of ions, whereas the situation of constant surface charge is generally assumed when isomorphic substitutions take place [85].

The LSA assumes that a region exists between the two interacting surfaces where the potential is sufficiently small and obeys the Poisson-Boltzmann equation, so that contributions from each surface can be added to give the overall potential. The LSA approximation for sphere-plate electrical double layer interaction energy ΔG^{EL} used in this work is described by the following equation [86]:

$$\Delta G^{EL}(x) = 64\pi\varepsilon\varepsilon_0 R \left(\frac{K_B T}{ze}\right)^2 \tanh\left(\frac{ze\psi_1}{4K_B T}\right)\tanh\left(\frac{ze\psi_2}{4K_B T}\right)\exp(-\kappa x) \tag{6}$$

where ε is the permittivity of free space, ε_0 the static dielectric constant of water, R the colloid radius, K_B the Boltzmann's constant, T the temperature, z the valence of the ion, e the electronic charge, Ψ_1 and Ψ_2 the surface potentials of both involved surfaces. κ is the inverse Debye length and x the separation distance between colloid and mineral surface.

2.2 Van der Waals interactions

A detailed understanding of van der Waals forces took a long time to emerge [91]. These forces may not be as strong as Coulomb or H-bonding interactions but they are the most ubiquitous bonding type since are always effective [92].

Calculation of the van der Waals force was initially deduced by Hamaker 1937 [87] from microscopic and later by Lifshitz 1956 [88] from macroscopic observations. Furthermore, a thermodynamic description of the interaction forces is possible which can be derived from the free adhesion energy (Johnson-Kendall-Roberts- [89] or Derjaguin-Müller-Toporov-theory [90]). However, only the microscopic model by Hamaker will be discussed here.

2.2.1 Microscopic model according to Hamaker

According to the microscopic theory of Hamaker, the energies of all single atoms and molecules in one body are integrated with all the atoms in the other body to yield the "two-body" potential for an atom near a surface, for a sphere near a surface or for two flat surfaces. This procedure can also be carried out for other geometries [87]. The van der Waals force includes three kinds of dipole-dipole interactions [92]:

- Keesom- or orientation-dependent interaction
- Debye- or induction interaction
- London- or dispersion interaction

Keesom- or orientation-dependent interaction

The Keesom interaction energy describes the interaction between two freely rotating permanent dipoles. They attract each other because they preferentially orient with the opposite charges facing each other. The orientation specific coefficient C_{orient} is independent of the distance. The Keesom-interaction energy w_{orient} (r) with the dipole moments μ_1 and μ_2 is described by the following equation [92]:

$$w_{orient}(r) = -\frac{\mu_1^2 \mu_2^2}{3(4\pi\varepsilon_0)^2 K_B T r^6} = -\frac{C_{orient}}{r^6} \qquad (7)$$

where μ_1 and μ_2 are the dipole moments of dipole 1 and 2 and r is the distance between atoms or molecules. C_{orient} is the orientation coefficient.

Debye- or induction interaction

Also a molecule with a static dipole moment can interact with a polarisable molecule. If the dipole can freely rotate, the dipole-induced dipole energy is: [91,92]:

$$w_{ind}(r) = -\frac{\mu^2 \alpha}{(4\pi\varepsilon_0)^2 r^6} = -\frac{C_{ind}}{r^6} \qquad (8)$$

where α is the polarizability of the molecule. C_{ind} is the induction coefficient.

London-or dispersion interaction

The London-or dispersion interaction is caused by alternately induced dipoles. The electron density around the nucleus is fluctuating in space and time. This leads to an undulation of the charge distribution around the nucleus and thus to a formation of fluctuating dipoles. The interaction energy $w_{disp}(r)$ for two different atoms or molecules 1 and 2 with distance r to each other can be described [91,92] by the following equation:

$$w_{disp}(r) = -\frac{3}{2}\frac{\alpha_{0,1}\alpha_{0,2}}{(4\pi\varepsilon_0)^2}\frac{h\nu_1\nu_2}{(\nu_1+\nu_2)r^6} = -\frac{C_{disp}}{r^6} \tag{9}$$

where $\alpha_{0,1}$ and $\alpha_{0,2}$ is the polarizability of the atoms and molecules, respectively; ν_1 and ν_2 are their ionization frequencies and h is Planck's constant. C_{disp} is the specific London interaction coefficient.

The van der Waals interaction energy $w_{vdW}(r)$ according to Hamaker is calculated by summing up the individual interaction contributions described above [91, 92]:

$$w_{vdW}(r) = -\frac{C_{vdW}}{r^6} = -\frac{C_{disp}+C_{orient}+C_{ind}}{r^6} \tag{10}$$

where C_{vdW} is the van der Waals coefficient.

The equations above show, that all terms contain the same distance dependency: the potential energy between molecules decreases with $1/r^6$. For macroscopic bodies the decay is less steep and depends on the specific shape of the interacting bodies [92].
However, often only the dispersion interaction is considered for the calculation of van der Waals forces, since this contribution always takes effect between all molecules and even between uncharged ones [91]. The assumptions made by Hamaker referring to the integration or summing up of all atoms (or molecules) of material 1 and a non-similar material 2 allow geometry-dependent equations, which always contain the Hamaker constant A_{H12}. The Hamaker constant is defined as [91]:

$$A_{H12} = \pi^2 \rho'_1 \rho'_2 C_{vdW} \tag{11}$$

where ρ'_1 and ρ'_2 are the number densities of the atoms or molecules.

If the Hamaker constant A_H of the material is known, it is possible to calculate the Hamaker-constant for the whole system in dependence of the involved solids 1, 2 and the surrounding medium 3 [91]:

Two solids 1 (A_{H11}) and 2 (A_{H22}) interacting with each other in vacuum:

$$A_{H12} \approx \sqrt{A_{H11} A_{H22}} \qquad (12)$$

A system with a solid 1 (A_{H11}) and 2 (A_{H22}) surrounded by fluid 3 (A_{H33}) [91]:

$$A_{H132} \approx \left(\sqrt{A_{H11}} - \sqrt{A_{H33}}\right)\left(\sqrt{A_{H22}} - \sqrt{A_{H33}}\right) \qquad (13)$$

The main setback of the Hamaker approach is that until now only few Hamaker constants were determined and the published values vary strongly. A reason for this are the different calculation- and experimental methods [93]. For example, for titanium dioxide, Hamaker-constants of $31 * 10^{-20}$ J [93] and $15.3 * 10^{-20}$ J [94] are given, which correlates to a deviation of factor two.

The indirect calculation of the Hamaker constant of a material is possible by the relatively simple determination of the surface energies. By measuring the contact angle of liquid drops of known surface tension on a solid surface it is possible to calculate the surface energy γ of the solids (for a liquid, γ is usually referred to as its surface tension) [95,96]. This method was also applied to determine the Hamaker constants of the samples used in this work.

Van der Waals interactions between a sphere and a plate were calculated using an approximate equation for the retarded van der Waals attraction energy ΔG^{vdW}, suggested by Gregory [97], which gives good agreement with exact solutions at short separations (up to 20 % of particle radius).

$$\Delta G^{vdW}(x) = -\frac{A_H R}{6x(1+(14x/\lambda_c))} \qquad (14)$$

Van der Waals forces are subject to a retardation effect due to their electromagnetic character. The finite time of propagation causes a reduced correlation between oscillations in the interacting bodies and a smaller interaction [81]. For spherical particles, it was shown that retardation of van der Waals forces is important at close approach. For instance, with spheres of 1 μm, separated by a distance of 10 nm, the attraction energy is reduced by a factor of about 2. At a distance of 100 nm, retardation causes about a 10 fold reduction in attraction. In equation 14, a retardation effect is introduced by the empirical factor $1/(1+14x/\lambda_c)$. It is convenient to think of a characteristic wavelength λ_c of the interaction which is given by $\lambda_c=2\pi c/\omega_v$ where c is the speed of light and ω_v the dispersion frequency. The characteristic wavelength has a value of 100 nm for most materials and is taken in this work [81]. The range of the van der Waals forces is about 10-15 nm [98].

2.3 Aggregation and stability of colloidal dispersions

The colloid-mediated transport of contaminants in subsurface environments is strongly dependent on the aggregation of the colloids where different interaction mechanisms have to be considered. For this purpose, it is important to describe the immediate surrounding of the colloid-water interface. The double layer of colloidal particles has been described by the following models [99].

The Helmholtz model compares the interface between electrolyte and mineral surface with a plate capacitor which stores charges. Counter ions bind directly to the surface and neutralize the surface charges. The disadvantage of this model is the lack of consideration of interactions beyond the first layer and the failure to explain the capacitance of an electric double layer [92].

The Gouy-Chapman model assumes infinite, flat impenetrable interfaces, ions in solution as point charges which are prone to thermal motion, able to approach right up to the interface, surface charge and potential are "smeared out" over the interface and the solvent is a uniform medium with properties that are independent of distance to the surface. The thermal fluctuations of the ions lead to the formation of a diffuse layer, which is more extended than a molecular layer. The Gouy-Chapman theory has several shortcomings. For instance, measured capacitances at certain interfaces can be much lower than those predicted by theory. Also, counterion concentrations close to charged interfaces can become unreasonably high, even for only moderate values of surface potential. Relatively simple modifications can be made, the most important of which are to allow for finite ion size and specific adsorption of ions. These considerations lead to the Stern and Stern-Grahame models of the double layer [81].

The Stern model considers ions in solution to have finite size. This is probably the most important correction to the Gouy-Chapman theory. Fig. 6 shows the different ion layers and the potential distribution in the immediate vicinity of a negatively charged colloidal particle according to the Stern model [81].

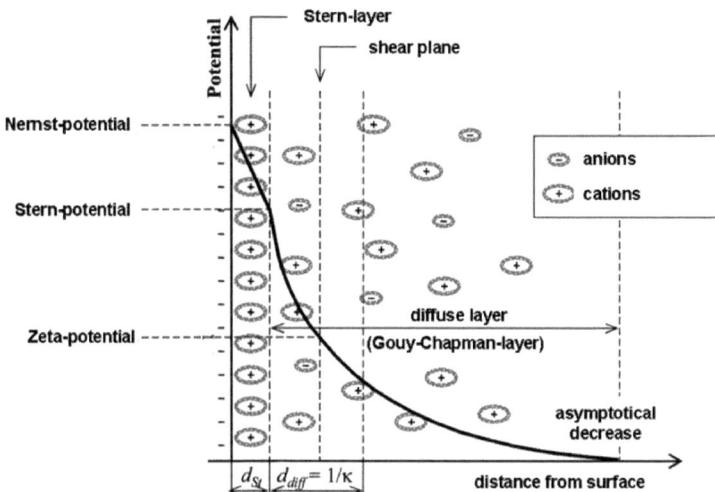

Fig. 6: Layers of ions and potential distribution in the immediate vicinity of a negatively charged colloid particle dispersed in electrolyte solution [101] (see text)

As already discussed above (see section II-2.1), most aquatic colloids possess a surface charge known as the Nernst-Potential. A monolayer of cations adsorbs onto the negatively charged particle surface, the so-called Stern layer. The ions in this layer are directly adsorbed onto the surface and are immobile [92]. By adsorption of counter ions the negative potential in the Stern layer decreases linearly to the Stern potential. The Stern layer is surrounded by a diffuse layer consisting of cations and anions and the concentration of anions increases with increasing distance from the particle surface. The diffuse layer is also known as the Gouy-Chapman layer. In contrast to the Stern layer, the ions in the diffuse layer are mobile and can be described by Poisson-Boltzmann statistics [92]. The potential at the point where the bound Stern layer ends and the mobile diffuse layer begins is the zeta potential. The negative potential in the diffuse layer decreases exponentially with increasing distance from the particle surface and falls asymptotically to zero. The characteristic length or "thickness" of the electric double layer is known as the Debye-length. The Debye-length d_{diff} is the reciprocal value of the Debye-Hückel parameter [92]:

$$d_{diff} = \frac{1}{\kappa} \qquad (15)$$

κ depends on the ionic strength and dielectric constant of the dispersion medium and can be from calculated the following equation [92]:

$$\kappa = \left(\frac{8\pi e^2 N_A I}{1000\varepsilon_0 K_B T} \right)^{\frac{1}{2}} \qquad (16)$$

where N_A is Avogadro's constant and I is the ionic strength.

In very dilute electrolyte solutions the diffuse layer is greatly expanded. With increasing electrolyte concentration, more counterions accumulate in the vicinity of the particle leading to a steeper decrease of the potential and a decrease of the Debye-length. This effect is known as "compression" of the diffuse layer.

Although the Stern Model is adequate for many purposes, a number of refinements can still be made. The Stern-Grahame model subdivides the Stern layer into two regions: an inner layer occupied by specifically adsorbed, unhydrated ions (e.g., inner sphere complexes) and a second layer where hydrated counterions (e.g., outer sphere complexes) are located. The boundaries of these regions are often known as the inner Helmholtz plane (IHP) and outer Helmholtz plane (OHP). The OHP is equivalent to the Stern plane. Outside this plane is the diffuse layer, described by the Gouy-Chapman theory. Fig. 7 shows a schematic diagram of the essential features of the Stern-Grahame double layer model. The adsorption of unhydrated ions at the IHP and of hydrated ions at the OHP, together with a diffuse layer which extends outwards into the bulk solution are shown. The electric shear boundary is thought to lie just outside the OHP. An important advantage of this model is that it recognises three different values of the solvent permittivity, e.g., the dielectric constant of the inner region, outer region and diffuse layer may have different values.

II Literature review 20

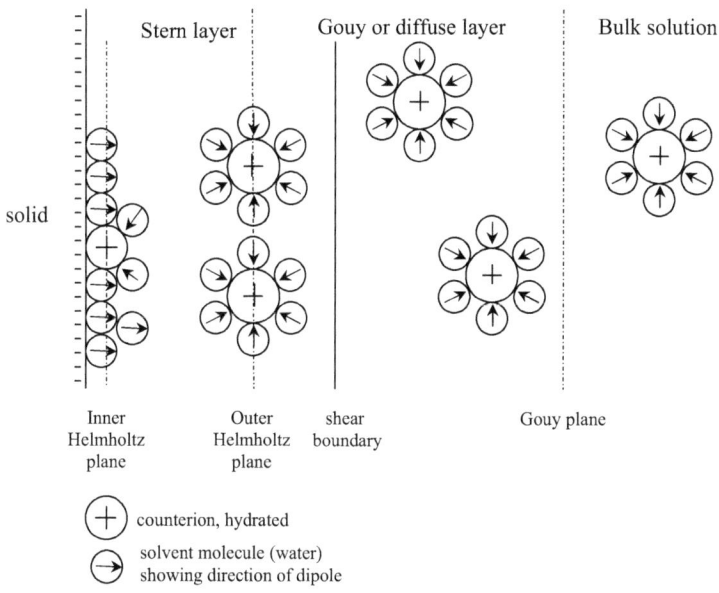

Fig. 7: The Stern-Grahame model of the electrical double layer (see text)

2.4 Van der Waals and double-layer forces acting together: the DLVO theory

Colloid dispersions are indicated as stable when a sedimentation of the particles can not be measured within a definite time (for example between 2 and 24 h) and the particle size distribution is not changing within this time span [100]. The stability of colloid dispersions depends mainly on electrostatic repulsion and attractive van der Waals forces. The influence of these forces on the colloid stability and particle-particle interactions is described by the DLVO-theory (DLVO: Derjaguin, Landau, Verwey, Overbek). The DLVO-theory considers the total interaction energy between similar charged particles which result from the energy of the electrostatic repulsion ΔG^{EL} and the van der Waals attraction ΔG^{vdW} as a function of distance between the similar charged particles (Fig. 8). In this Figure also the non-DLVO Born forces are also shown - these will be discussed in section II-2.5. In dilute electrolyte solutions electrostatic repulsion dominates at middle and high particle distances (Fig. 8 (a)). The total interaction energy ΔG^T which is the sum of electrostatic double layer repulsion and van der Waals energy, leads at middle distances to an energy barrier ΔG^{max}. Is this energy barrier multiple times higher than the thermal energy $K_B T$, the particles will not overcome

ΔG^{max} and the colloidal dispersion remains stable. In case of ΔG^{max} being not high enough, the particles may overcome the energy barrier and aggregate. At even smaller distances the Born repulsion ΔG^B dominates. The total interaction energy is then $\Delta G^B + \Delta G^T$. The Born repulsion prevents any further approach and leads to a so-called primary minimum. The aggregates formed in the primary minimum are usually dense. At very high particle distances the van der Waals attraction may dominate the electrostatic double layer repulsion and a secondary minimum is formed. In contrast to the primary minimum the aggregates formed are here relatively loosely bound and voluminous. Because of the comparatively weak attractive forces in the secondary minimum the aggregates may be redispersed, for example by generating shear forces through stirring [101].

In media containing electrolytes with higher concentration the electrostatic double layer forces are often weaker and van der Waals forces dominate over the whole particle-particle distance. As a consequence, the total interaction energy lies mainly in the attractive regime (Fig. 8 (b)) and the particles aggregate.

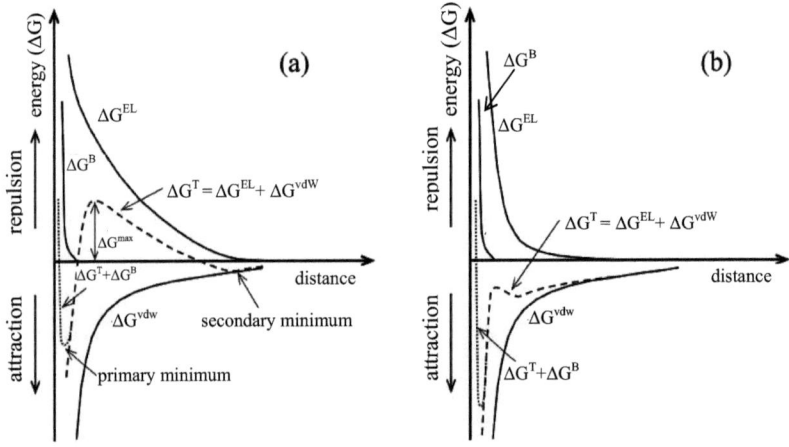

Fig. 8: Electrostatic repulsion ΔG^{EL}, van der Waals attraction ΔG^{vdW} and Born repulsion ΔG^B energies and their total ΔG^T as a function of distance between similar charged particles dispersed in electrolyte at low (a) and high (b) electrolyte concentration [102] (see text)

For a number of colloidal systems it is found that the critical ion concentration necessary for colloid aggregation varies with the inverse sixth power of the valency of the electrolyte counterions. It is a empirical observation, known as Schultz-Hardy rule [91]:

$$[Me^+] \div [Me^{2+}] \div [Me^{3+}] \equiv \left(\frac{1}{1}\right)^6 \div \left(\frac{1}{2}\right)^6 \div \left(\frac{1}{3}\right)^6 \qquad (17)$$

where [Me] is the metal cation concentration.

The stability of colloidal dispersions is also usually influenced by the pH. Colloids with surface functional groups such as hydroxyl groups, react very sensitively to pH changes. Through protonation or deprotonation of these groups the surface charge can change in dependence of the pH. At the pH_{pzc}, the repulsive forces are minimal and attraction results.

2.5 Non-DLVO forces

There are situations, in which the DLVO theory does not give satisfactory agreement with experimental results. Assuming that the DLVO theory is reliable in itself, in such cases some kind of additional short-range forces may have to be considered. For cases in which the precise details of the spatial variation of the short-range forces are not important, it is often more convenient to resort to the microscopically averaged Born repulsion [81]. For colloidal particles carrying adsorbed polymers, the forces are known as steric or osmotic forces. Structural forces are associated with the forces that develop as particles with adsorbed fluid layers interact. These forces describe the interactions arising specifically from the adsorption of solvent, surfactant or macromolecules at the interface. It covers the repulsive hydration interactions, the attractive hydrophobic interactions and the repulsive steric interactions [91]. In the following section a selection of relevant non-DLVO forces is described.

Born repulsion

This short-range repulsion originates from the strong repulsive forces between atoms as their electron shells interpenetrate each other. Only quantum mechanical considerations are able to give a precise description of the interatomic potential. However, a number of simplified approximate analytical forms have been proposed. A relatively realistic formulation, at least at the approximation level of the DLVO theory, is the Hamaker-type integration of all molecules in the system [91]. Ruckenstein and Prieve derived a formula obtained in this manner for the Born repulsion involving the interaction of a sphere with a plate [103]:

$$\Delta G_{Born}(x) = \frac{A_H \sigma_c^6}{7560}\left[\frac{8R+x}{(2R+7)^7} + \frac{6R-x}{x^7}\right] \qquad (18)$$

II Literature review

where the collision diameter σ_c is typically in the order of 0.5 nm. The effect of Born repulsion may not be of great significance in aqueous systems since the presence of any hydrated ions, which are likely to be present, will prevent surface-surface separation distances approaching 0.3 nm [91].

Hydration forces

The hydration force is not of a simple nature and is probably the most important yet the least understood of all the forces in liquids [91].
The nature of water close to a surface can be for a number of reasons very different from that of bulk water. Since most surfaces have a surface charge and hence ionic surface groups, some hydration of these groups is expected. The approach of two particles with hydrated surfaces will generally be hindered by an extra repulsive interaction, distinct from the electrical double layer repulsion. When two hydrated surfaces are brought into contact repulsive forces of about 3-5 nm range have been measured in water on clays, mica, silica, alumina, lipids and DNA [91]. Because of the correlation with the low (or negative) energy of wetting of these solids with water, the repulsive (hydration) force has been attributed to the energy required to remove the water of hydration from the surface, or surface adsorbed species, presumably because of strong charge-dipole, dipole-dipole or H-bonding interactions. It is assumed that several effects contribute to the short-range repulsion and a broader discussion can be found in the literature [81,91,92]. When reviewing literature about AFM force measurements in aqueous medium many papers do not mention hydration forces; most are concerned with long-range forces such as the electrostatic double-layer and the van der Waals attraction. The fact that only few papers deal with hydration forces is probably due to the problems involved. Problems arise due to deformation of the sample colloid probe, the precise determination of zero distance, and contamination, which can cause a steric repulsion [92].

2.6 Hamaker constants and contact angle measurements

Van der Waals interaction forces, the surface energy, the work of adhesion and the Hamaker constants all are related to one another, which will be explained in the following text.
Literature values of surface energies differ even for the same characterisation method, i.e., contact angle measurements (e.g., compare measurements of [104] and [105]). Often surface contamination, surface roughness, applied dropsizes and different interpretation methods are

the reasons for this. A significantly higher discrepancy can be found in the literature regarding the Hamaker-constants. Hamaker constants can be derived from the dielectric functions for all frequencies of the solids, gas chromatography or contact angle measurements. To measure the dielectric function over the whole frequency range, different spectroscopic methods like electron energy loss spectroscopy, UV, infrared and microwave spectroscopy must be applied [183]. In this work, the Hamaker constants are determined by the derivation of surface energies obtained from contact angle measurements. This is a consistent method to determine the Hamaker constants and allows the comparsion of values obtained in different studies.

2.6.1 Three-phase equilibrium and the Young-Dupré equation

A liquid drop on an ideal smooth solid surface surrounded by the gas phase shows by its contact angle the conditions at the three-phase contact line [106] (Fig. 9). If the contact angle θ is measured while the volume of the drop is increasing (this is done just before the wetting line starts to advance) the advancing contact angle is observed. When afterwards the volume of the drop is decreased and the contact angle is determined just before the wetting line is receding, the so-called receding contact angle is observed. The difference between both angles is called contact angle hysteresis [92] caused for example by surface roughness or chemical heterogeneities.

The advancing contact angle of a drop of liquid can be treated as force balance, as it represents the resultant of the energy of cohesion of the liquid vs. the energy of adhesion between liquid and solid. Only the advancing contact angle can fulfil the physical requirements of force balance, because only an advancing contact angle guarantees that the line at the liquid-solid-air interface is situated on a part of the solid surface that has not been previously wetted by the liquid of the drop. This type of measurement is therefore the most reproducible way of measuring contact angles. As a result, advancing angles are normally measured in order to determine the surface free energy of a solid. The receding angle is not suitable for calculating surface energies [108].

The three-phase equilibrium between the surface energy of the solid γ_S, the surface tension of the liquid γ_L and the solid-liquid interface γ_{SL} is described by the Young equation:

$$\gamma_L \cos\theta = \gamma_S - \gamma_{SL} \tag{19}$$

The smaller the contact angle, the better is the wetting of the surface. Low values of θ indicate a strong liquid-solid interaction such that the liquid tends to spread on the solid, or wets well, while high θ values indicate weak interaction and poor wetting. For θ < 90° the liquid is said to wet the solid. A zero contact angle represents complete wetting. For θ > 90°, the liquid is non-wetting. [107].

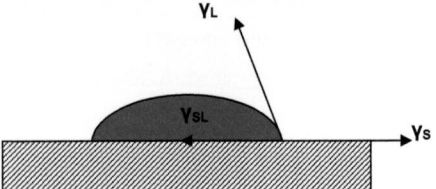

Fig. 9: Formation of a contact angle between a liquid drop at the three phase liquid–solid–gas phase interface. θ is the angle between γ_L and γ_L (see text)

2.6.2 Interpretation of the contact angle measurements

Besides the measurable contact angle θ the interfacial tension between the liquid and the solid γ_{SL} is unknown and has to be determined by approximation procedures. Basis and starting point of the determination of surface energies is always Young's equation. In this work, only the Lifshitz-van der Waals acid-base approach according to van Oss et al. [108] was used and will be described here.

Van Oss and Good differentiate between a polar (acid-base interactions) and an apolar (Lifshitz-van der Waals) component of the surface energy (γ^{LW}). The polar interactions are defined as interactions resulting from the sharing of electrons (or protons) between surface functional groups and other polar molecules (or protons). In contrast to other methods (i.e., equation of state [109], Zisman-method [110]) van Oss and Good describe the polar fraction with the help of the Lewis acid-base model. According to this model, the polar fraction of the surface energy of the solid and the surrounding drop liquid is split into an electron acceptor fraction corresponding to a Lewis acid (= electron acceptor fraction γ^+) and an electron donor corresponding to a Lewis base (= electron donor fraction γ^-). The presence or absence of γ^+ and γ^- groups on sample surfaces, in relation to the magnitude of the Lifshitz-van der Waals component, determines the surface's affinity for water (i.e., its hydrophobicity/hydrophilicity) [108]. A mineral surface with a large polar surface free energy, in relation to its Lifshitz-van der Waals component, tends to be hydrophilic, while a surface characterized by a small acid-base component tends to be hydrophobic. On hydrophilic surfaces, water structuring is induced as water molecules form hydrogen bonds with surface γ^+ and γ^- groups. Thus, when two hydrophilic surfaces are brought into contact, a hydrophilic repulsion results and energy must be expended to remove the adsorbed water layers [108].

Taking these different surface tension components into account, the Young's equation becomes known as the Young-Dupré equation [108]:

$$(1+\cos\theta)\gamma_L = 2\left(\sqrt{\gamma_M^{LW}\gamma_L^{LW}} + \sqrt{\gamma_M^+\gamma_L^-} + \sqrt{\gamma_M^-\gamma_L^+}\right) \qquad (20)$$

The subscripts L and M stand for liquid and mineral surface, respectively. In order to solve this equation, i.e. to determine the disperse fraction γ^{LW}, the acid fraction γ^+ and the base fraction γ^- of the solid, contact angle data from at least three test liquids are required; at least two of these must have known acid and base fractions > 0.

Moreover, at least one of the liquids must have equal basic and polar parts. Usually water is chosen for this purpose because it serves as neutral point in the Lewis scale.

The values of liquid surface tension components are readily available from the literature [108], and the advancing contact angle is determined experimentally.

2.6.3 Interrelation between surface energy and Hamaker constant

The work of adhesion per unit area W_{12} is the free energy change which is required to separate two media 1 and 2 from contact to infinity in vacuum. If the materials are identical this energy is referred to as the work of cohesion. It is [92]:

$$W_{12} = \gamma_{11} + \gamma_{22} - \gamma_{12} \qquad (21)$$

where γ_{12} is the surface energy of the interface of material 1 and 2 and γ_{11} and γ_{22} the surface energies of the pure materials. In order to create a new surface of a unit area, two half unit areas have to be separated from contact to infinity [92]:

$$\gamma_{12} = \frac{W_{12}}{2} \qquad (22)$$

For the estimation of the surface energy γ_{12} by using the surface energies of the pure materials γ_{11} and γ_{22} the following equation can be applied [91]:

$$\gamma_{12} = \gamma_{11} + \gamma_{22} - 2\sqrt{\gamma_{11}\gamma_{22}} = \left(\sqrt{\gamma_{11}} - \sqrt{\gamma_{22}}\right)^2 \qquad (23)$$

Implementing also van der Waals interactions (see eqn. 10) the interaction energy can be calculated, i.e. the work of adhesion between two infinitely large surfaces is [91]:

II Literature review

$$W_{12} = \frac{\pi C_{vdW} \rho'_1 \rho'_2}{12x^2} \tag{24}$$

By taking the definition of the Hamaker constant (see eqn. 11) and equalising Eqn. 22 with eqn. 24 the surface energy is interrelated with the Hamaker constant as can be seen by the following equation [91]:

$$W_{12} = \frac{A_{H12}}{12\pi x^2} = 2\gamma_{12} \tag{25}$$

It is not obvious what value to use for the interfacial contact separation x because in the Hamaker summation method the discrete surface atoms are artificially "smeared out" into a continuum by transferring the energy sum to an integral. By assuming that the surfaces are made up of atoms or molecules in most dense sphere packing, x can be estimated as x ≈ $a_0/2.5$. A typical value of the interatomic or intermolecular centre-to-centre distance a_0 is 0.4 nm; x is then about 0.165 nm. Equation 26 describes how the Hamaker constant can be estimated when the surface energy is known [91]:

$$A_{H11} = 24\pi(0.165)^2 \gamma_{11} \tag{26}$$

According to the approach based on the surface free energy by van Oss, the Hamaker constant is calculated by using the disperse fraction γ^{LW} of the surface tension [108]:

$$A_H = 24\pi(0.165)^2 \gamma^{LW} \tag{27}$$

The above equations can be applied individually to mineral surfaces, colloid and water to derive A_{HM}, $A_{Hcolloid}$, A_{Hwater} from $\gamma_M^{LW}, \gamma_{colloid}^{LW}, \gamma_{water}^{LW}$ which are in turn derived from contact angle measurements [108].

III Methods and Materials

1 Methods

1.1 Determination of zeta-potential

1.1.1 Electrophoresis

Electrophoresis was used to determine the zeta-potential of natural and synthetic colloids. When an electrical field is applied to a suspension, charged particles begin to move to the oppositely charged electrode. Consequently, positively charged particles move to the cathode and negatively charged particles to the anode. This movement of charged particles in an electrical field is called electrophoresis. The higher the particle charge, the higher is also the particle velocity and thus the so-called electrophoretic mobility. The electrophoretic mobility μ_E is the quotient of particle velocity v_p and field strength E:

$$\mu_e = \frac{v_p}{E} \tag{28}$$

To calculate the zeta-potential ζ from the electrophoretic mobility the Henry-equation is used [20]:

$$\zeta = \frac{3\mu_e \eta}{2\varepsilon f(\kappa R)} \tag{29}$$

where η is the dynamic viscosity of the dispersing medium and f(κR) the Henry-function. For particles > 20 nm and an electrolyte concentration above 1 mmol/l f(κR) is 1.5. In this case, the Smoluchowski-equation is applicable [81]:

$$\zeta = \frac{\mu_e \eta}{\varepsilon} \tag{30}$$

For smaller particles which are dispersed in very dilute electrolyte solutions, the Hückel-approximation is used (f(κR) = 1) [81]:

$$\zeta = \frac{3\mu_e \eta}{2\varepsilon} \tag{31}$$

III Methods and Materials

The light scattering behaviour of the particles moving in the electric field is the basis of determining the zeta-potential. Hereby different methods such as Laser Doppler Anemometry (LDA) and Phase Analysis Light Scattering (PALS) are used. In this work only PALS is applied. In PALS, the phase of the laserlight is modulated in such a way that the Doppler frequency of the scattered light from the moving particles is equal to the modulated frequency. Even if the particles have a very small mobility, the phase shifts and the smallest phase change can be detected by a phase comparator. The average phase shift is described by the following equation [111]:

$$\langle Q(t) - Q(0) \rangle = \langle A_s \rangle K [\langle \mu_e \rangle E(t) + u_c] \qquad (32)$$

with the amplitude-weighted phase at the time t Q(t) and 0 Q(0), the signal amplitude A_s, the scattering vector K, the field strength E(t) and the velocity of the collective movement u_c (e.g. because of the temperature gradient). Zeta-potential measurements of the colloids were performed with a Zeta Plus (Zeta Potential Analyser, Brookhaven Instruments Corporation, USA). Each zeta-potential value was an average of 10 runs each with 30 measurements. The zeta-potential was calculated using the von Smoluchowski theory [112].

1.1.2 Streaming Potential

This method was used to characterize surface electrical properties such as the zeta-potential of mineral surfaces. Liquid is pressed through a cylindrical tube with a charged inner wall. In consequence, the liquid drags the charges of the electrical double layer with it. The counterions accumulate at the end of the tube and generate an electrical potential difference between the ends of the tube, which is the streaming potential. The streaming potential causes a conduction current which just balances the streaming current [92]. When the Debye length is very small compared to the tube radius the streaming potential can be calculated by

$$\Delta U = -\frac{\varepsilon_0 \varepsilon \zeta}{\eta \kappa_e} \Delta P \qquad (33)$$

where ΔP is the pressure difference between the ends of the tube and κ_e is the electrical conductivity of the electrolye. The streaming potential experiments in this work were performed with a Surpass Electrokinetic Analyzer (Anton Paar GmbH, Graz, Austria).

1.2 Photon Correlation Spectroscopy (PCS)

This method uses dynamic light scattering and is able to detect colloidal particles with hydrodynamic diameters of 1-6000 nm. Besides the detection of the average colloid size the size distribution of particles can be measured.

When light interacts with matter scattering of the incident beam occurs. If the matter is homogenous, the scattered light interferes destructive and nearly total extinction of the light takes place. If the matter is inhomogencous (e.g., colloidal dispersions, macromolecular solutions, aerosols and smoke) only partial extinction takes place and the observer perceives scattered light. In 1871, the light scattering of colloid dispersions was mentioned for the first time by Tyndall [20]. The scattering behaviour of particles can be described by the ratio of particle diameter d and the wavelength λ of the light by the following theoretical approximations: Rayleigh-approximation $(d < \lambda/10)$, Rayleigh-Gans-Debay-approximation $(\lambda/10 < d < \lambda/2)$, Mie-theory $(d \approx \lambda)$ and the Fraunhofer-approximation $(d > 5\lambda)$.

For colloidal particles, the Rayleigh-approximation is of relevance (particle diameter is below $\lambda/10$). The theory of dynamic light scattering is based upon the principle of Brownian diffusion of the colloidal particles. The diffusion coefficient of the colloids and their size distribution is determined by time-resolved detection of the scattered light. Regarding the light scattering in colloidal dispersions, the phase shift leads to constructive and destructive interferences of the scattered light. The average scattered light intensity is the sum of the scattered light intensities of all individual particles [113]. Fig. 10 shows a scheme of the PCS.

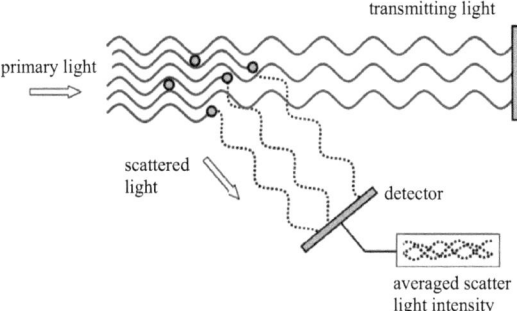

Fig. 10: Light scattering in a colloid dispersion: due to the phase shifts the scattered light interferes and this leads to amplification or extinction of the scattered light (see text)

Assuming spherical particles, the scattered light intensity I_s can be described by the following equation:

III Methods and Materials

$$I_s = \frac{I_0 \pi^4 \rho_d^2 d^6 \sin^2 \theta_a}{9\lambda^4 b^2} \left(\frac{dn'}{dc_M}\right)^2 \qquad (34)$$

where I_0 is the intensity of the incidental beam, ρ_d the density of the particles, d the particle diameter, θ_a the scattering angle, b the distance from the scattering centre, n the refraction index of the solution and c_M the mass concentration.

The detected summarized signal of the scattered light contains the information about the particle size. Due to the Brownian diffusion it comes to fluctuations of intensity. Since smaller particles have a higher diffusion coefficient and are more mobile than larger particles, the scattering signal fluctuates faster than that of bigger particles. The scattering signal is analysed by the auto correlation function. Thereby scattered light intensities are compared with each other in very small time increments (nano- and microseconds). The auto correlation function is described by the following normalized equation which is true for monodisperse samples:

$$q(\tau) = 1 + e^{-2DK^2\tau} \qquad (35)$$

where τ is the correlation time, D the diffusion coefficient of the particles, and K the scattering vector.

By using the diffusion coefficient D it is possible to calculate the hydrodynamic radius of the particles with the Stokes-Einstein-Equation:

$$r_p = \frac{K_B T}{6\pi\eta D} \qquad (36)$$

Advantages of dynamic light scattering are short measurement times, uncomplicated sample preparation, small sample volumes and the lack of necessity to calibration.
PCS measurements of the colloids were performed with the similar apparatus as the zeta-potential measurements.

III Methods and Materials   32

1.3 Contact Angle Measurements

Contact angle measurements were carried out to determine the material surface energy. This method requires homogeneous and preferably smooth sample surfaces, since sample roughness can strongly influence the macroscopically measurable contact angle [114] (see section II-2.6).

The contact angle measurements were carried out by depositing three selected test liquids on a prepared sample surface. The contact angle measuring apparatus (OCA 5, DataPhysics Instruments GmbH Filderstadt, Germany) consists of a manually adjustable sample table, a dosage system and a light microscope. The light source was positioned behind the drop, which then appears dark. The contact angle was determined directly with a goniometer. Advancing contact angles were measured. To solve for the three unknowns of the Young-Dupré equation (see section II-2.6.2), three independent liquids were used: dimethyl sulphoxide (DMSO), toluene and water with pH 6 and 2 (adjusted with 10^{-1} M NaOH and HCl). The ionic strength of the water was set to 10^{-2} M NaCl. The values of the surface tension components of the liquids were provided by van Oss [108] and are listed in Table 2. Before undertaking the experiments, the sample surfaces were immersed and washed with chloroform, methanol and MQ-water. Afterwards, the samples were irradiated with an UV-lamp to remove organic residues (lamp power 400 W). At least 20 advancing contact angles were measured per sample. The measurements were carried out on the mineral surfaces of quartz, muscovite, biotite and apatite.

The analysis software SCA10 (DataPhysics Instruments GmbH, Filderstadt, Germany) was used to calculate the surface energies with the Young-Dupré equation from the measured contact angles.

Table 2: Surface tension components in [mN/m] of the liquids used for the contact angle measurements. Values were taken from van Oss [108]

surface tension component	dimethyl sulphoxide	toluene	water
γ^{LW}	36	28,3	21,8
γ^+	3,12	0	25,5
γ^-	24	2,7	25,5

1.4 Fluorescence Microscopy

A fluorescence microscope is basically a conventional light microscope with adapted features and components. A scheme is shown in Fig. 11.

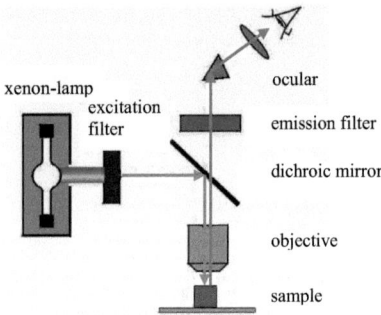

Fig. 11: Scheme of a fluorescence microscope (see text)

In fluorescence microscopy, the light illuminating the sample is used to excite fluorescing species on a sample, which then emit light with a longer wavelength. The fluorescence microscope also produces a magnified image of the sample, but the image is based on the light emanating from the fluorescent species rather than from the light originally used to illuminate and excite the sample. The fluorescence microscope uses the objective lens to focus the excitation light onto the sample and to collect the emitted fluorescence. The illumination (excitation) light is separated from the fluorescence emission emanating from the sample. A dichroic mirror is used to separate the excitation and emission light paths. In order to select the appropriate excitation wavelength, an excitation filter is placed in the excitation path just prior to the dichroic mirror. The excitation light reflects off the surface of the dichroic mirror into the objective and the fluorescence emission then passes through the dichroic mirror into the eyepiece.

A dichroic mirror has a set wavelength value - called the transition wavelength value - which is the wavelength of 50% transmission. The mirror reflects wavelengths of light below the transition wavelength value and transmits wavelengths above this value. Ideally, the wavelength of the dichroic mirror is chosen to be between the wavelengths used for excitation and emission. When the excitation light illuminates the sample, a small amount of excitation light is reflected off the optical elements within the objective and some excitation light is scattered back into the objective by the sample. Some of this "excitation" light is transmitted through the dichroic mirror along with the longer wavelength light emitted by the sample. This "contaminating" light would reach the detection system if an emission filter was not placed in the excitation path just prior to the dichroic mirror.

The fluorescence microscope used was a Leica DMIRE2 (Leica Microsystems GmbH, Wetzlar, Germany) with a filter set optimized for the fluorescence characteristics of the rhodamine-labeled colloids used in this work (λ_{Ex} = 552 nm, λ_{Em} = 580 nm) and a monochrome CCD-Camera (Lu135M, Lumenera Corporation, North Andover, MA, USA). Fig. 12 shows the transmission of the individual filters vs. wavelength of the filterset used.

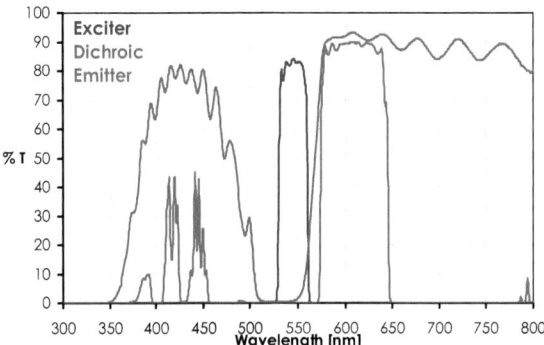

Fig. 12: Transmission vs. wavelength spectra of the exciter/emitter filters and the dichroic mirror (see text)

The fluorescence intensity (I_F) in fluids can be described by the following relation [115], which involves the Lambert-Beer Law:

$$I_F = \Phi_F I_0 (1 - e^{-\varepsilon_x c d}) \quad (37)$$

where ε_x is the extinction coefficient, d_{op} the optical pathlength and c is the concentration of the fluorophore, Φ_F the fluorescence quantum yield and I_0 the intensity of the exciting light. The quantum yield Φ_F is the relation between emitted and absorbed light quanta. For dry samples various parameters influence I_F, i.e., sample surface geometry or varying lamp intensities.

The fluorescence intensity was quantified by reading out mean grey-scale values for a selected image area (WinSpec/32 software, Version 2.5.19.7, Roper Scientific) and is given as counts in an arbitrary scale [a.u.].

These values range from 0-255 (8-bit tiff-format), with 255 being the most intense signal. A value of 17±0.3 counts was determined as an average background signal; this value was subtracted from the experimental signal values. The background signal was not influenced by the presence of metal cations, i.e., Eu(III) used in our experiments. The stability of the fluorescence signal over time periods of several weeks is controlled and corrected by using a fluorescent glass as reference standard (AHF Analysentechnik AG, Tübingen, Germany) before each measurement. The photochemical stability of the rhodamine-labeled latex colloids was investigated by continuous irradiation of a sample for four hours in the fluorescence microscope. No significant decrease of fluorescence intensity was detected. This experimental approach allowed a semi-quantitative comparison of fluorescence intensities derived from different samples or sample sites over longer periods of time. At least 50 fluorescence images of the surface were recorded for each sample. The obtained images had a size of 348×256 µm and the size of one pixel was ≈ 1 µm². For a better presentation, fluorescence-optical images were processed in order to enhance image contrast.

1.5 Scanning Electron Microscopy (SEM)

Scanning electron microscopes are routinely used to obtain a view of surface structures with an excellent lateral resolution of typically 1-20 nm. Furthermore, the high depth of field makes it possible to acquire a three-dimensional impression of the sample topography.

In the scanning electron microscope, an electron beam is emitted from a heated filament or a field emission tip. The electrons are accelerated by an electric potential in the order of 1-400 kV. A condenser lens projects the image of the source onto the aperture. The beam is focused by an objective lens and raster-scanned by scanning coils over the sample. The minimization of the spot size is important, since it determines the resolution of the instrument. Fig. 13 shows a scheme of the SEM. When the primary electrons hit the sample surface, they pass part of their energy to electrons in the sample, resulting in the emission of secondary electrons. These secondary electrons have low energies (≈ 20 eV) and therefore only those secondary electrons escape from the sample, which are close (≈ 1 nm) to the sample surface. The secondary electrons are collected by a detector and their intensity is displayed versus the position of the primary beam on the sample [92].

Besides the secondary electrons that are detected for the topographic imaging, there are also elastically backscattered primary electrons and X-rays generated by the interaction of the primary electrons with the sample atoms. The backscattering probability depends on the mean atomic number of the material. This imaging mode allows detection of material contrast but is less surface sensitive since the penetration depth of the backscattered electrons is in the range of some 100 nm. The X-ray emission can be used for X-ray spectroscopy of the sample that allows semi-quantitative analysis of the sample composition. If an electron from an inner shell is removed by X-rays or primary electrons, an electron from an outer shell fills the hole. During this process either a photon with an energy that corresponds to the difference between the two involved energy levels is emitted or this energy is transferred to another outer electron. This outer electron is then emitted which is called Auger electron. It is possible to analyze the energy of the emitted X-rays with the energy dispersive X-ray analysis (EDX) [92].

A CS44FE Field Emission SEM (CamScan, Cambridge, UK) was used in this work.

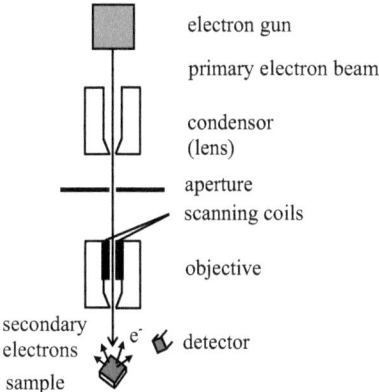

Fig. 13: Scheme of the SEM [92] (see text)

1.6 Atomic Force Microscopy (AFM)

In the 1980s a series of scanning probe techniques were developed, e.g., the scanning tunnelling microscopy [116,117], scanning near-field optical microscopy [118,119] and the AFM, invented by Binning and coworkers [120].

In AFM, a very fine tip attached to a cantilever is scanned over the sample surface; the distal end of the tip is usually in mechanical contact with the sample. The apex of the tip has a radius of about few tens of nanometers. The cantilever is moved by a piezoelectric scanner in the xy-plane and the cantilever deflection is measured very precisely.

III Methods and Materials   37

There have been a number of detection modes for the cantilever deflection reported in the literature, including the measurement of the tunnel current between cantilever and a reference electrode [121], capacitive [122], interferometric [123] and piezoresistive [124] detection methods. The most commonly applied method used in modern instruments is the so-called optical lever technique [125,126,127]. A scheme is shown in Fig. 14.

In the optical lever technique, a laser beam is focused onto the rear side of the cantilever to detect the movement of the tip. The reflected beam irradiates a photodetector with at least two segments. The position of the reflected beam changes with the movement of the cantilever and the photodetector converts this variation into an electrical signal. The lateral resolution is not limited by a fundamental length of interaction such as the wavelength of light, but rather by the geometry and distance of the probe. A resolution from submicrometer to atomic dimensions can be achieved. Besides the actual measuring unit, which consists of the detection system with the cantilever, laser diode and photodetector as well as the piezoscanner, the sample is placed onto a vibration-isolated table and there are also the feedback and scan control system with a computer which belong to the basic components of the AFM. While the computer generates the control signals for the movement of the sample, the scan control system processes and amplifies these signals.

The AFM used in this work is a Dimension 3100 Digital Instruments (Santa Barbara, USA).

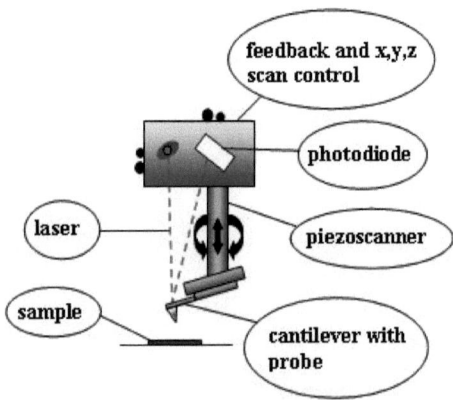

Fig. 14: Scheme of the AFM (see text)

1.6.1 Cantilever and detection

The cantilever is the actual sensor of the AFM and thus a key component. Important for the force measurements is the knowledge of the spring constant and the deflection behaviour, which affect the detection sensitivity. For the scanning process and imaging a preferably hard and sharp tip is desirable in order to avoid tip wear and to maximize the spatial resolution. Today mostly photolithographic silicon [128] and silicon nitride cantilevers [129] are used. On the rear side of the cantilevers a gold or aluminium layer is applied to increase the reflexion properties of the optical lever method.

For the AFM measurements, cantilevers with different specifications, depending on the measuring mode, were used, as shown in Table 3.

Table 3: Specifications of cantilevers used in this work for imaging

	tapping mode	contact mode
tip height:	15 – 20 µm	2.5 - 3.5µm
tip radius (nominal):	10 nm	20 nm
tip radius (max.)	12.5 nm	60 nm
material:	phosphorus (n) doped Si	silicon nitride
nominal spring constant of cantilever	40 N/m (nom.)	0.06 N/m (nom.)
nominal cantilever length	125 µm	196 µm
nominal cantilever width	35 µm	23 µm

1.6.2 Measuring modi

Contact mode

When the AFM is operated in contact mode the probe is always in mechanical contact with the sample surface and hence within the area of influence of the short-range interaction forces. Consequently atomic resolution can be achieved. The load force is in the range of 10^{-9} and 10^{-7} N. A limitation of the contact mode is that sensitive samples can be damaged during the scanning due to shear forces. For this reason in contact mode, generally soft cantilevers with spring constants of about 0.1 N/m are used. To reduce tip wear, mostly cantilevers consisting of silicon nitride are employed. The normal force signal is taken as a measure for the cantilever deflection and thus for the height deflection of the tip relative to the sample. Scanning in contact mode can be regulated or unregulated. In the regulated contact mode ("constant force mode") the normal force signal is held constant by using an electronic feedback-loop by application of adequate voltages onto the piezoelectric scanner. The

III Methods and Materials   39

feedback signal as a function of the tip location is used for the generation of images with topographical information [130].

In the unregulated contact mode ("constant height mode") the distance between cantilever and sample is held constant and the normal force signal is directly recorded. If the elastic deformation of the sample surface is disregarded, then the sample topography can be derived from the height deflection of the cantilever.

Non-contact mode

In non-contact mode, the tip is scanned over the sample surface with a separation of 10 to 100 Å and the cantilever is excited to oscillate with small amplitudes (in the range of a few nanometers). The excitation frequency is chosen close to the resonance frequency. Thereby the tip is influenced by the long-range, mostly attractive interaction forces. Their local gradient perpendicular to the sample surface leads to a change of the effective spring constant of the cantilever and hence to a shift of the resonance frequency. At unchanged excitation frequency a shift of oscillation amplitude and phase results from this. During the measurement, the oscillation amplitude is held constant by controlling the distance between cantilever and sample. This feedback control signal is recorded as in the regulated contact mode. Besides this method, other feedback modi exist in which the excitation frequency varies and the frequency shift is recorded [131,132].

Intermittent contact mode

The intermittent contact mode (also called tapping mode) operates in a similar fashion as the non-contact mode [133]. However, higher oscillation amplitudes are used here (in the range of 10 to 100 nm) [134] and the cantilever is drawn so close to the sample surface that the tip touches the sample periodically. The oscillation amplitude is limited through strong contact interaction and is held constant during scanning with the feedback-loop. This feedback control signal is used to generate images with topographic information. The phase shift between excitation oscillation and cantilever oscillation is also a measure for the dissipated energy in the tip-sample system [135]. The Intermittent-contact mode is widely used also because the comparatively small shear forces generated allow imaging of sensitive samples.

1.6.3 AFM force spectroscopy

The development of the theory of van der Waals forces stimulated an interest in measuring forces between surfaces to verify this theory [136]. With the AFM, surface forces can be measured directly and relatively universally by obtaining so-called force-distance curves. The force is measured between a sample surface and a microfabricated tip which is placed at the end of a ca. 100 µm long and 0.4-10 µm thick cantilever. Alternatively, colloidal particles are fixed on the cantilever; this technique is called the "colloid probe technique". Using this method, the forces between surfaces and colloidal particles can be also directly measured in liquid [137,138]. The Fig. 15 shows a colloid probe cantilever used in the experiments of this work.

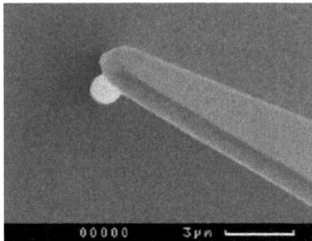

Fig. 15: AFM colloid probe cantilever: the colloid particle (white sphere) is 1 µm in diameter (see text)

The specifications of the cantilevers used in the force spectroscopy experiments are shown in Table 4.

Table 4: Specifications of cantilevers used in the force spectroscopy experiments

property	
colloid size	1 µm (nom.)
cantilever material	silicon nitride
spring constant of cantilever	0.12 N/m (nom.)
cantilever length (µm)	196 (nom.)
cantilever width	23 (nom.)

In contrast to the measuring modes discussed in section III-1.6.2, AFM force spectroscopy is not an imaging scanning mode. The normal force signal as a function of distance between cantilever and sample surface is measured on a chosen sample location. As a result, force-distance curves can be obtained: during the recording of the force-distance curves the z-position of the cantilever relative to the steady sample is measured. The normal force signal is hence a measure for the normal deflection of the cantilever.

The direct result of a force measurement is a measure of the photodiode current versus height position of the piezoelectric translator. To obtain a force-distance curve, the photodiode current and piezo position have to be converted into force and distance. For this, two parameters have to be known: the deflection sensitivity and point of zero distance. Both parameters must be inferred from the force curve itself and not through an independent method. The linear part of the "contact regime" is assumed to be the zero distance and its slope the deflection sensitivity [92].

Sample runs of force-distance measurements between a carboxylated latex sphere and a muscovite surface at different pH are pictured in Fig. 16 (measured at pH 2, I = 10^{-2} M NaCl) and Fig. 17 (measured at pH 10, I = 10^{-2} M NaCl). At a certain distance, the interaction forces are initially negligible and the cantilever is in its force free neutral position. On approaching the surface, attractive interaction forces may increase, causing the spring constant of the cantilever to be overcome. This leads to an instability and the tip jumps onto the sample ("snap-in"; Fig. 16). As soon as the tip is in contact with the sample surface, the adhesion forces are compensated by a short-ranged repulsive contact interaction. This interaction prevents a further penetration of the tip into the sample, so that the tip follows the sample surface with further approach. During the retraction of the cantilever, the tip stays due to adhesion even further in contact with the sample until finally the spring constant becomes stronger than the adhesion force and a second instability occurs: the tip jumps away from the surface. The force measured here is a so-called "snap-off" force and is used to determine the adhesion force. In the most general case, the adhesion force F_{ad} is a combination of the electrostatic force F_{el}, the van der Waals force F_{vdW}, the capillary force F_{cap} (only in air) and forces due to chemical bonds or acid-base interactions F_{chem} [92]:

$$F_{ad} = F_{el} + F_{vdW} + F_{cap} + F_{chem} \tag{38}$$

Depending on the chemical end-groups on tip and substrate, chemical bonds may form during contact or other specific interactions (e.g. receptor – ligand) may occur and then often dominate the adhesion force [92]. The negative peaks of the approach and retraction curves are taken as values for snap-in and adhesion forces, respectively.

The difference in the path between approach and withdrawal curves is usually called "force-distance curve hysteresis".

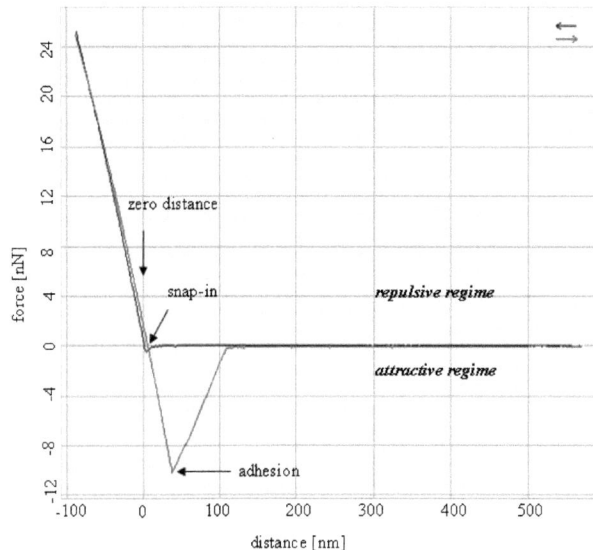

Fig. 16: Force-distance curve between a carboxylated latex sphere and a muscovite surface (pH 2, I = 10^{-2} M HCl): the blue line represents approach of the cantilever, the red line retraction (see text)

The experimental force-distance curve measured on muscovite at pH 10 (Fig. 17) shows that no attractive forces were evident but the repulsive electrostatic double layer force is apparent before the cantilever comes into contact with the mineral surface.

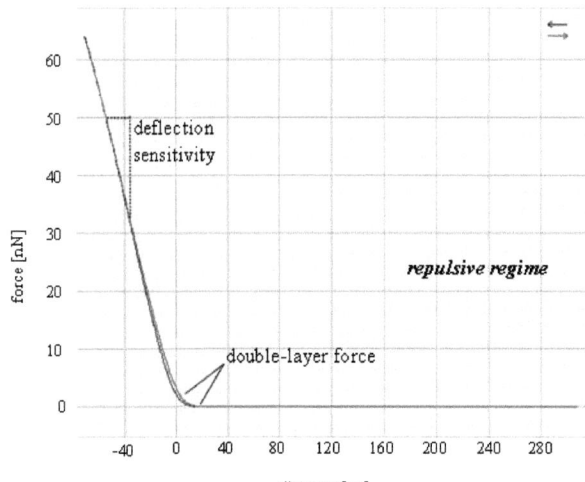

Fig. 17: Force-distance curve between a carboxylated latex sphere and a muscovite surface (pH = 10 I = 10^{-2} M NaCl). The blue line represents approach of the cantilever, the red line retraction (see text)

III Methods and Materials

The measured force-distance curves were analysed with SPIP© analysis software (Image Metrology, Hørsholm, Denmark).

Calibration of cantilevers

Quantitative force measurements depend on the accurate determination of the cantilever spring constant. Although the cantilever deflection x_d can be measured with great accuracy and sub-Angstrom sensitivity, converting these measurements to units of force via Hooke's law, ($F = -k_s x_d$) requires that the spring constant, k_s, be determined for each cantilever. It has often been noted in the literature that spring constants can vary greatly from the values quoted by their manufacturers. In fact, these values are only provided as nominal indications of the cantilever properties and the manufacturers often specify the spring constant in a wide range that may span values up to four times smaller and four times larger than the nominal value [139].

Various techniques exist for the calibration of cantilever spring constants. These can be generally grouped into three categories: (a) "Dimensional models" where fully theoretical analysis or semi-empirical formulas are used to calculate the cantilever spring constants based on their dimensions and material properties [140,141,142,143,144]; (b) "Static deflection measurements" determine the spring constant by loading the cantilever with a known static force. (c) "Dynamic deflection measurements" measure the resonance behaviour of the cantilever and relate it back to its spring constant [145,146,147,148,149,150]. In this work, static deflection measurements of the cantilevers are undertaken to determine the spring constant. These methods are based on the premise that the spring constant can be measured by applying a known force on the cantilever and measuring its deflection. The static deflection method applied in this work was to use a calibrated reference cantilever [151]. Here the cantilever with unknown spring constant measures force curves on the end of a second cantilever that is calibrated ("reference cantilever"), whereas the reference cantilever can be fixed on a, e.g., glass slide by double sided tape. The (average) slope of the contact portion of the obtained force curves (i.e., the deflection sensitivity in nm/V) is compared to that measured on a hard surface (e.g., the glass slide) and the spring constant k_s is calculated from:

$$k_s = k_{ref}\left(\frac{S_{ref}}{S_{hard}} - 1\right) \tag{39}$$

III Methods and Materials   44

where k_{ref} is the spring constant of the reference cantilever, S_{ref} is the deflection sensitivity measured on the reference cantilever, and S_{hard} is the deflection sensitivity measured on a hard surface.

Reference cantilevers with very well controlled dimensions and material properties (i.e., the spring constant) were acquired from Veeco Instruments GmbH, Mannheim, Germany. The uncertainty of the reference cantilever method is about 9 % and the main source of error is the measurement of deflection sensitivity [152].

Force-volume Measurements

By using AFM force-volume imaging it is possible to measure a two-dimensional array of regularly spaced force curves on a sample area. Each force curve is measured at a unique x-y position in the area, and force curves from an array x-y points are combined into a three-dimensional array or "volume", of force data. The value at a point (x,y,z) in the volume is the deflection (force) of the cantilever at that position in space. Force volumes allow investigation of the spatial distribution of almost any force between tip (colloid) and sample that varies with the distance between the two. It is possible to acquire 2-dimensional maps depicting the snap-in or the adhesion forces within a selected sample area. Force-volume images of the sample forces were recorded with several mineral surfaces in order to evaluate heterogeneities of the measured snap-in and adhesion forces on the sample areas. Experimental conditions were pH 5.8, I = 10^{-2} M NaCl; the Ca(II) concentration was set to 10^{-4} M. The experiments were carried out with the same colloid probe cantilevers as used in the single-point measurements The scan area in these measurements was set to 15 x 15 µm and the resolution was 16 x 16 measuring points, so that the whole measurement area was probed.

Calculation of maximum potential energy barriers

Colloid-collector attachment efficiency (the probability that colloids adsorb onto collector surfaces despite the presence of a repulsive energy barrier) is a function of solution chemistry and the surface properties of the particle and collector. It was shown, that the attachment efficiency is proportional to $\exp(-|\Phi|/K_B T)$ [153], where Φ is the surface potential, K_B is Boltzmann's constant and T is the absolute temperature. The proportionality factor can be viewed as a frequency factor and depends on some physical and colloidal properties of the deposition system [154]. The ratio $\Phi/K_B T$ is a dimensionless form of the maximum potential interaction energy. The higher the maximum potential energy barrier, the smaller is the probability of colloid attachment.

The maximum potential energy barrier values were calculated as a function of the intersurface potential energy (Φ) between a particle and collector. In this study, force (approach) curves calculated by DLVO theory have been integrated over separation distance to calculate Φ directly:

$$energy(\Phi) = \int_{0.3}^{x} F\,dx \qquad (40)$$

The integration is performed from the first point of interaction (x) detected by the AFM until the particle and collector make contact (separation distance 0.3 nm).

In this work, the calculation of the maximum potential energy barriers is undertaken on selected force spectroscopy measurements with the software Origin 8G SR2 (OriginLab Corporation, Northhampton, USA).

2 Materials

2.1 Colloids

For the colloid sorption experiments, fluorescent carboxylate-modified polystyrene colloids (functionalized during polymerization) with a diameter of 25 nm were purchased from Postnova Analytics (Landsberg/Lech, Germany). The nanospheres were rhodamine-labeled with an excitation wavelength of 552 nm and an emission wavelength of 580 nm. The concentration of the stock suspension was 10 g/l. The polydispersity index was provided by the manufacturer as <0.1 and the density of COOH groups is 0.12 mmol/g. The pH of the stock suspension was 4.3. Sample solutions were prepared by diluting the colloid stock suspension with 10^{-2} M NaCl in Milli-Q water to a final colloid concentration of 0.05 g/l.

Natural FEBEX-Bentonite (Cabo de Gata deposit, Almeria, Spain) was sieved and equilibrated with 1 M NaCl to transfer the bentonite to its mono-ionic Na-form. The Na exchanged bentonite was washed with Milli-Q water to remove excess of salt after one week equilibration time. The suspension was centrifuged at 4000 rpm for 40 min. Afterwards the supernatant was discarded and the precipitate was resuspended in Milli-Q water. This washing cycle was repeated three times. Each washing step was checked by electrical conductivity measurements; the electrical conductivity after all washing cycles was 1.9 µS/cm. The

extracted colloidal fraction consisted quantitatively of montmorillonite (proved by XRD, Bruker AXS D8-Advance). The bentonite colloids used for our experiments have an average diameter of 250 nm determined by PCS [155].
Additionally, carboxylate-modified polystyrene microspheres (1 μm diameter) were acquired from Polysciences Inc. (Warrington, Pennsylvania, USA). The stock suspension concentration was 25 g/l. Similar latex microspheres were glued onto the AFM cantilevers by Novascan Technologies, Ames, Iowa, USA. Only zeta-potential measurements were carried out with these colloids (see section IV-1.2). For these measurements, the corresponding colloid suspension was washed with MQ-water and centrifuged several times to remove detergents. No information about the COOH density of these microspheres could be given from the manufacturer.

2.2 Sample solutions

Teflon bottles were used to prepare the sample solutions in order to avoid adsorption of colloids or cations on the bottle material itself. The ionic strength was held constant at 10^{-2} M NaCl in the sorption and force spectroscopy experiments. All experiments with background electrolyte only or in presence of Ca(II) were carried out from pH 2-10. The Ca(II) concentration was set to 10^{-4} M; this corresponds to the Ca(II) concentration in Grimsel groundwater. All experiments > pH 6 were carried out in an Ar-atmosphere to prevent CO_2 absorption by the sample solutions. From pH 2-10 Ca(II) exists as free aquo-ion, as proven by theoretical calculations with ECOSAT 4.8 (Wageningen University, Wageningen, Netherlands).
All experiments in the presence of Eu(III) (CertiPUR Europium ICP Standard, Merck, Germany) were undertaken in the pH range of 2-6 only, since Eu(III) undergoes hydrolysis and may precipitate under alkaline conditions [156]. Eu(III) is used as chemical homologue for trivalent actinides. Under the conditions of our experiments, the Eu(III)-aquo ion exists in solution and interacts with the colloid or the mineral surfaces. The Eu(III) concentration in the sorption and force spectroscopy experiments was set to 10^{-5} and 10^{-6} M.
Additional experiments with UO_2^{2+} were carried out because Uranium is a main component of spent nuclear fuel rods and occurs under environmental conditions as mobile UO_2^{2+} ions. Experiments in the presence of 10^{-6} M UO_2^{2+} were carried out in the pH range of 2-6. The speciation of UO_2^{2+} in water is very complex, since various hydrolysis species (dependent on pH, ionic strength and UO_2^{2+} concentration) exist. Fig. 18 shows the speciation of UO_2^{2+} in aqueous solution calculated also with ECOSAT 4.8. It can be seen that under the conditions of

our experiments the species mainly present species are UO_2^{2+}, $(UO_2(OH))^+$, $((UO_2)_3(OH)_5)^+$, $(UO_2(OH_3))^-$ and $(UO_2(OH)_2)$. More information about the speciation of UO_2^{2+} can be found in [157].

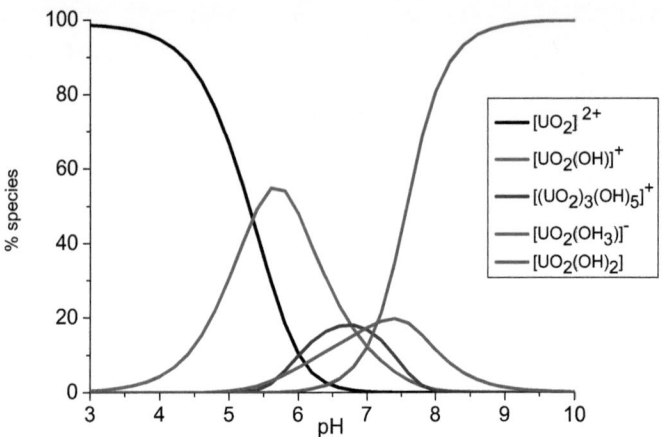

Fig. 18: Speciation of UO_2^{2+} in aqueous solution, in absence of CO_2. $[UO_2^{2+}] = 10^{-6}$ M, I = 10^{-2} M NaCl [157] (see text)

Force spectroscopy and sorption experiments were also carried out with natural Grimsel groundwater to compare measured values obtained from synthetic waters with the natural groundwater. The element concentration in the Grimsel groundwater was determined by ICP-MS (Perkin-Elmer ELAN 6000). As can be seen, the main elements present in the groundwater are Na, Si and Ca (Table 5). The ionic strength of the Grimsel groundwater is about 10^{-3} M, the pH is 9.6 [158]. In the colloid sorption experiments with Grimsel groundwater, the colloid concentration was set to 0.05 g/l.

Table 5: Element concentration of Grimsel groundwater

Element	concentration [mol/l]
Li	$1.05*10^{-5}$
Na	$4.01*10^{-4}$
Mg	$3.18*10^{-6}$
Al	$6.97*10^{-7}$
Si	$1.87*10^{-4}$
K	$5.26*10^{-6}$
Ca	$1.27*10^{-4}$
Mn	$1.16*10^{-6}$
Ni	$4.09*10^{-8}$
Cu	$2.20*10^{-8}$
Zn	$2.22*10^{-6}$
Rb	$1.99*10^{-8}$
Sr	$1.58*10^{-6}$

3 Sorption and force spectroscopy experiments

3.1 Sorption experiments with Grimsel granodiorite

The individual Grimsel granodiorite samples were equilibrated for three years in Grimsel groundwater (pH 9.6, I = 10^{-3} M; [8]).

For the sorption experiments using fluorescent latex colloids, a piece of Grimsel granodiorite (2×3 cm) was withdrawn from the groundwater and rinsed thoroughly with MQ-water. Then the granodiorite was dipped into the colloid suspension (volume 3 ml). Colloids were allowed to adsorb from fluid to sample surface for 15 min (T = 22 °C). This sorption time was chosen because sorption experiments with longer sorption times up to 2 hours showed no increase in colloid adsorption. The samples were then withdrawn from the suspension and thoroughly rinsed with MQ-water in order to remove any non-adsorbed latex colloids. After drying in an oven at 40°C, the rock samples were investigated by fluorescence microscopy. SEM/EDX was used to identify the minerals on which a predominant colloid adsorption took place.

To gain insight into the desorption of colloids from Grimsel granodiorite, an additional long-term desorption experiment taking 37 days was carried out. The colloids were adsorbed onto the granodiorite for 30 min. at pH 4, I = 10^{-2} M NaCl. The colloid concentration was set to 0.05 g/l. These conditions were chosen to ensure maximum colloid adsorption while minimizing mineral dissolution, which generally increases with decreasing pH values. The Grimsel granodiorite sample was then placed in an electrolyte solution (pH 10 and I = 10^{-2} M

III Methods and Materials   49

NaCl). The sample was withdrawn after different time intervals and the fluorescence intensity was measured after drying. The electrolyte solution was exchanged after each measurement to ensure that the rock sample interacts with colloid-free solution. Again SEM/EDX was used for identification of the mineral phases.

3.2 Single minerals used in sorption experiments

All single minerals (biotite, albite, K-feldspar, apatite, quartz and titanite) were commercially available (Krantz Mineralien Kontor, Bonn, Germany). Muscovite was purchased from Plano GmbH (Wetzlar, Germany) and sapphire from Mateck GmbH (Jülich, Germany). With the exception of sapphire, all these minerals are present in the Grimsel granodiorite. Additional experiments with sapphire (001 plane) were carried out since this mineral can be regarded in some respects as a model for clay minerals and analogous iron phases [159]. First, mineral surfaces were cleaned to remove any organic fluorescing impurities on the mineral surface. The minerals were immersed in and rinsed with chloroform, methanol and MQ-water [160]. After this procedure the samples were placed in an oven at 40°C for approximately 3 hours. Sheet silicates like muscovite or biotite were freshly cleaved in order to obtain clean surfaces. Sorption experiments were carried out as described for the granodiorite (see section III-3.1).

SEM/EDX was applied to determine the elemental composition of the mineral phases. The samples were sputtered with a thin chromium layer and investigated with a relatively low acceleration voltage of 15 kV in order to minimize the penetration depth of the electron beam. On each sample, several areas (size $200 \times 300 \mu m$) were analysed. The averaged values of the elemental compositions (in atom %) are summarized in Table 6. The error of the EDX measurements is estimated to 0.5 - 1.1 atom-%, depending on sample matrix and acceleration voltage.

Table 6: EDX data for the single minerals used in the sorption experiments (elemental composition in atom %, values < 1 % not included)

Element	albite	orthoclase	biotite	muscovite	quartz	apatite	titanite	sapphire
Si	21	22	8	10	30		12	
Al	8	7	4	8				43
Na	7							
Mg			4					
K		6	2	3				
Ca						23	15	
P						13		
Fe			2					
O	61	62	63	63	66	61	57	58
Ti							15	

3.3 Single minerals used in force spectroscopy experiments

Force spectroscopy measurements were carried out on the relevant minerals present in the Grimsel granodiorite (muscovite, biotite, quartz, K-feldspar, apatite, titanite). Surface roughness measurements of the minerals were undertaken in contact mode (in air) directly after the cleaning procedure. The scanning area was 25 µm². The RMS roughness, defined as the root-mean-square of all the distances from the center line of the roughness profile, calculated over the profile length, was obtained using SPIP© analysis software (Image Metrology, Hørsholm, Denmark). The measured surface roughness of the minerals is shown in Table 7.

Table 7: Mineral surface roughness determined by AFM

mineral	RMS roughness [nm]
muscovite	< 0.1 - 0.3
biotite	0.4 – 1.4
quartz	3.4 – 4.2
K-feldspar	1.9 - 3.2
apatite	2.2 - 3.4
titanite	6.1 – 7.2

SEM/EDX (CS44FE Field Emission SEM, CamScan, Cambridge, UK) was applied to determine the average elemental composition of the minerals used for the force spectroscopy experiments. On each sample, several areas (size 200 x 300 µm) were analysed. The results are summarized in Table 8.

Table 8: EDX data for the single minerals (elemental composition in atom %, values < 1 % not included) used in the force-spectroscopy experiments

Element	K-feldspar	biotite	muscovite	quartz	apatite	titanite
Si	25	8	10	30		12
Al	8	4	8			
Na	1					
Mg		4				
K	8	2	3			
Ca					26	15
P					17	
Fe		2				
O	58	63	63	66	50	57
F					6	
Ti						15

IV Results

1. Zeta-potential and PCS measurements

Zeta-potential measurements were carried out on colloids and mineral surfaces to gain insight into their charge properties under different geochemical conditions. Knowledge of the zeta-potentials of the colloids and mineral surfaces is necessary to understand the interaction behaviour of the colloids with the mineral surfaces, e.g., whether the conditions are generally repulsive or attractive.

PCS measurements were undertaken to ascertain the colloid stability behaviour under the selected experimental conditions. It was important to use conditions under which the colloids were stable and their agglomeration and sedimentation could be excluded.

As stated above, Eu(III) is used in the experiments as chemical homologue for trivalent actinides. Experiments with 10^{-4} M Ca(II) were carried out because Ca(II) is the dominating cation in Grimsel groundwater.

1.1 Fluorescent polystyrene colloids

Charge properties of the fluorescent carboxylated polystyrene colloids used in the sorption experiments were characterized over the pH range of 2-10 by zeta-potential measurements with the ionic strength set to 10^{-3} and 10^{-2} M NaCl and a colloid concentration of 1 g/l.

IV Results

Colloid suspensions with an ionic strength of 10^{-3} M NaCl were only measured in the pH range of 3-10 (lower pH values can not be adjusted at this ionic strength).
As shown in Fig. 19, the colloids were negatively charged in the observed pH and ionic strength range with zeta-potential values ranging from -50 to -20 mV depending on pH and ionic strength. Additionally, zeta-potential measurements in the presence of 10^{-5} M Eu(III) were measured with an ionic strength of 10^{-2} M NaCl and pH values ranging from 2-6. No significant influence of Eu(III) on the measured zeta-potentials was detected (not shown). As can be seen, the latex colloids are negatively charged over the whole pH range. The zeta-potentials of natural bentonite colloids with an average diameter of 250 nm were determined with an ionic strength of 10^{-2} M NaCl. The zeta-potential values of the natural bentonite colloids were significantly lower in the observed pH range than those determined for the latex colloids (Fig. 19).

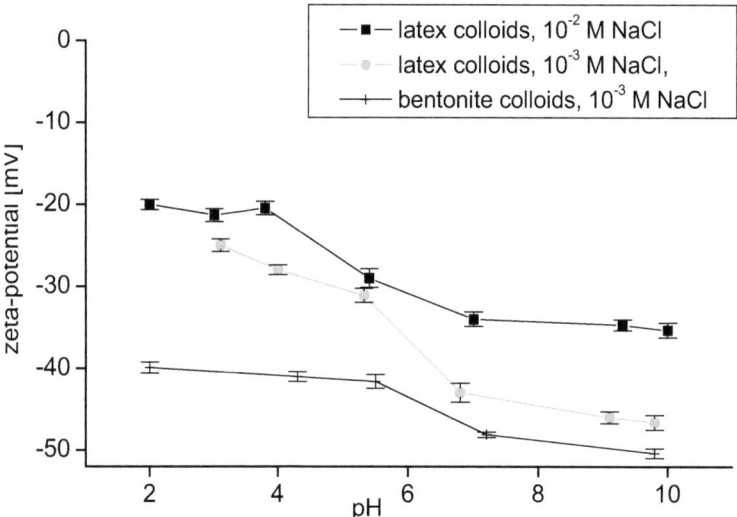

Fig. 19: Zeta-potential measurements of fluorescent polystyrene colloids and bentonite (see text)

In order to investigate the agglomeration behaviour of the colloids, PCS measurements were carried out in presence/absence of 10^{-5} M Eu(III). For these measurements pH values were varied from 2-6 and ionic strength was held constant at 10^{-2} M NaCl (colloid concentration 0.05 g/l). A colloid diameter of 25 ± 3 nm was determined over the whole pH range and even in the presence of Eu(III) in a time period of approximately 30 min. Therefore, it can be concluded that colloids are stable under the chemical conditions adjusted in the sorption

IV Results

experiments. Neither PCS nor zeta-potential measurements provided an indication that Eu(III) is adsorbed on the colloids under the given chemical conditions. From the literature it is known that Eu(III) complexation with carboxylated polymers is not significant at pH < 4 [161].

1.2 Polystyrene spheres used as AFM colloid probes

Zeta-potential measurements were also carried out with the carboxylated polystyrene microspheres (1 µm diameter). Experiments were performed with background electrolyte only (10^{-2} M NaCl) and background electrolyte with different cation concentrations ($10^{-5}/10^{-6}$ M Eu(III) and 10^{-4} M Ca(II)). Fig. 20 shows that the polystyrene colloids are negatively charged over the whole pH range and that the zeta-potentials follow a similar trend compared to the to the natural bentonite colloids. The natural colloids generally had a more negative charge than the 25 nm fluorescent polystyrene colloids (compare Fig. 19 with Fig. 20). In contrast to the measurements shown in Fig. 19, a clear impact of Eu(III) on the zeta-potentials was observed in the experiments with the 1 µm particles. A possible reason for this are the generally lower (more negative) zeta-potentials of the 1 µm particles (compare black squares in Fig. 19 and Fig. 20).

IV Results 54

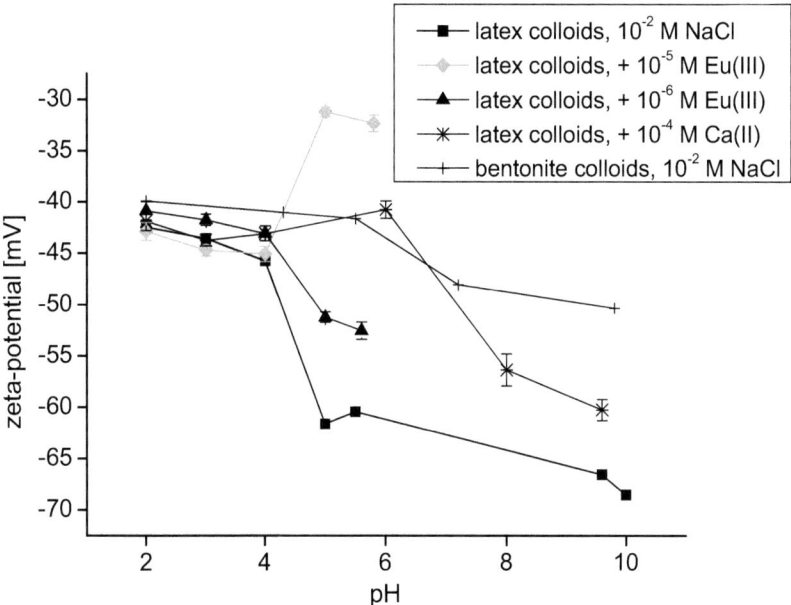

Fig. 20: Zeta-potentials of the carboxylated latex colloids as used in force spectroscopy and bentonite (see text)

1.3 Streaming potential measurements

Streaming potential measurements were carried out with quartz to investigate its zeta-potential in dependence of pH. Due to the limitations of this method (a constant sample surface geometry during the measurement is required) it was only possible to carry out the measurements on quartz.

Experiments were undertaken with 10^{-2} M NaCl background electrolyte and additionally also in presence of 10^{-5} M Eu(III). The experiments in absence of Eu(III) showed that the pH_{pzc} lies approximately at pH 3.8. Above this value, the zeta-potential decreases with increasing pH to values of -60 mV. The experiments in the presence of Eu(III) showed a shift in the pH_{pzc} to approximately 4.3. The zeta-potentials above this value increase with increasing pH due to the increasing adsorption of Eu(III) onto the quartz surface; this behaviour is similar to the increase of zeta-potential due to the Eu(III) adsorption onto the 1 µm polystyrene particles (compare red squares in Fig. 21 with grey diamonds in Fig. 20).

IV Results 55

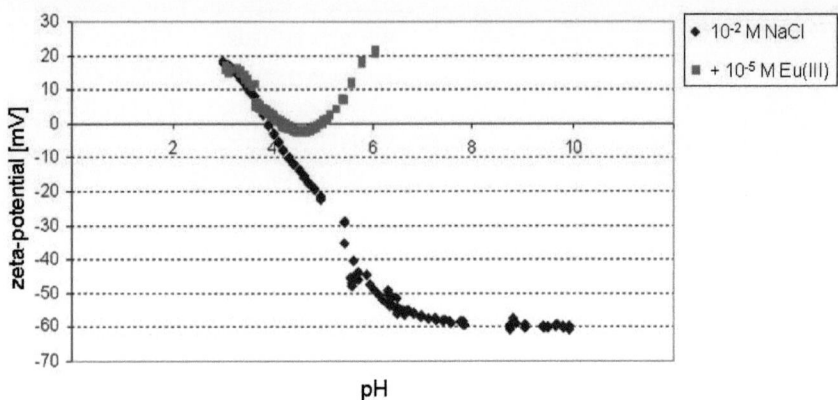

Fig. 21: Zeta-potential of quartz in presence and absence of Eu(III) (see text)

2. Sorption experiments with fluorescent polystyrene colloids

Sorption experiments under varied geochemical conditions with fluorescent polystyrene colloids and Grimsel granodiorite and its component minerals were carried out in the initial phase of this work. The aim was to detect adsorbed colloids on the natural surfaces by fluorescence microscopy. By measuring the fluorescence intensity, a semi-quantitative measure of the adsorbed colloids on the individual mineral surfaces (identified by SEM/EDX) could be achieved. The measured fluorescence intensities were also correlated with a colloid surface coverage on the individual minerals.

2.1 Surface coverage measurements

The colloid surface coverage on the individual mineral surfaces was determined by correlating fluorescence intensities of adsorbed colloids on muscovite (measured by fluorescence microscopy) with the number of colloids per unit area (determined by AFM). Sorption of colloids on muscovite was used here as reference system because adsorbed colloids could be easily detected on the atomically flat surface. Colloid concentration was varied in the range 0.5-2 g/l in order to find a sufficient number of colloids on a selected image area. Eu(III) was added with a concentration of 10^{-5} M to increase the colloid adsorption. Images were obtained by tapping-mode AFM on dried samples. The number of colloids per μm^2 was counted at several image areas with different fluorescence intensities. An image of latex colloids adsorbed onto muscovite is shown in Fig. 22. The height of the colloids in Fig. 22 was determined as ca. 25 nm and the diameter was in the range of 50-

100 nm. It is well-known that lateral dimensions are strongly overestimated due to the common AFM tip artefact, i.e. when the size of a feature is in the range of the tip radius. This is the case for the colloidal particles. The artefacts can be corrected by geometrical models. In the literature, various methods are proposed for the correction of tip artefacts [162,163]. The PCS measurements also did not show any coagulation of the colloids (see section IV-1.1). Therefore, it can be concluded that mainly single colloids on muscovite were found. Some aggregation on the surface, however, can not be excluded.

The area (A) occupied by a single spherical colloid was calculated by $A = \pi r_{nom}^2$, where r_{nom} is the nominal colloid radius (12.5 nm). The surface coverage was obtained summing up values of A and relating this sum to the total image area. Surface coverage is given in percent. A fairly linear relation between surface coverage and the fluorescence intensity is found. From the standard deviation (3σ) of the fluorescent background signal (17 ± 0.3 counts) the detection limit of the method was estimated to around 2 colloids/μm^2 corresponding to a surface coverage of around 0.1 %.

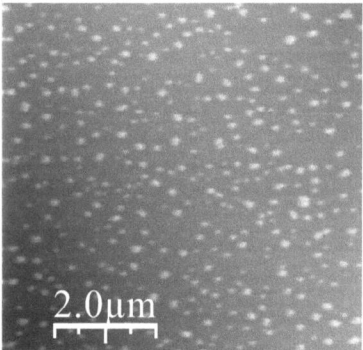

Fig. 22: AFM image of muscovite surface with adsorbed latex colloids (pH = 4, I = 10^{-2} M NaCl, $c_{colloids}$ = 2 g/l, Eu(III) = 10^{-5} M) (see text)

2.2 Interaction of carboxylated latex colloids with component minerals of Grimsel granodiorite

Single minerals relevant for the Grimsel granodiorite system were contacted with various suspensions containing fluorescent carboxylated latex colloids at different pH and Eu(III) concentrations (see section III-2.2). Additional experiments were carried out with Grimsel groundwater also containing the carboxylated latex colloids. In the fluorescence images, different regions with characteristic features can be distinguished. Sorption on surface areas with mineral edges was indicated, when fluorescence was detected along straight or wound lines. Sorption on the mineral planes was evident, when homogeneous fluorescence intensity

was detected on a certain area without a characteristic morphology. Further experiments were carried out with sapphire which, although, is not a component mineral of Grimsel granodiorite, is regarded in some studies as a model for minerals [159].
The results of all colloid sorption experiments are discussed in section V.

2.2.1 Sheet silicates (muscovite and biotite)

Fig. 23 (a) and (b) show fluorescence-optical images of muscovite surfaces after adsorption of fluorescent polystyrene colloids at pH 4 in absence (a) and presence (b) of 10^{-5} M Eu(III). The black coloured areas in Fig. 23 (a) represent the basal planes where measured intensity values were in the range of the background signal and, thus, no adsorption could be detected. In the bright areas increased fluorescence intensities were measured indicating colloid adsorption. Fluorescence signals could only be detected along straight lines or similar features which are typical for mineral edges. Fig. 23 (b) shows that in presence of Eu(III) colloid sorption increased and took also place on the muscovite basal planes.

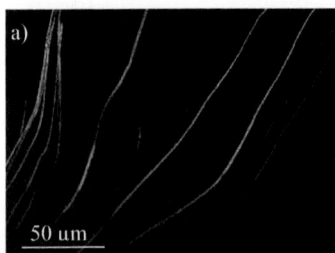
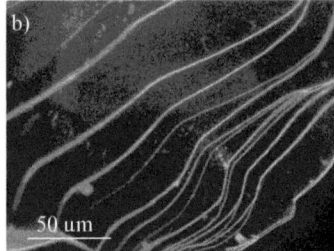

Fig. 23: Fluorescence-optical image of muscovite (a) after adsorption of fluorescent colloids (pH = 4, I = 10^{-2} M NaCl, $c_{colloids}$ = 0.05 g/l) and (b) in the presence of Eu(III) (pH = 4, I = 10^{-2} M NaCl, $c_{colloids}$ = 0.05 g/l, Eu(III) = 10^{-5} M) (see text)

Fig. 24 shows fluorescence intensity/surface coverage vs. pH obtained from the adsorption of carboxylated colloids on muscovite in the presence/absence of Eu(III). Fig. 25 shows similar data obtained for biotite.

IV Results 58

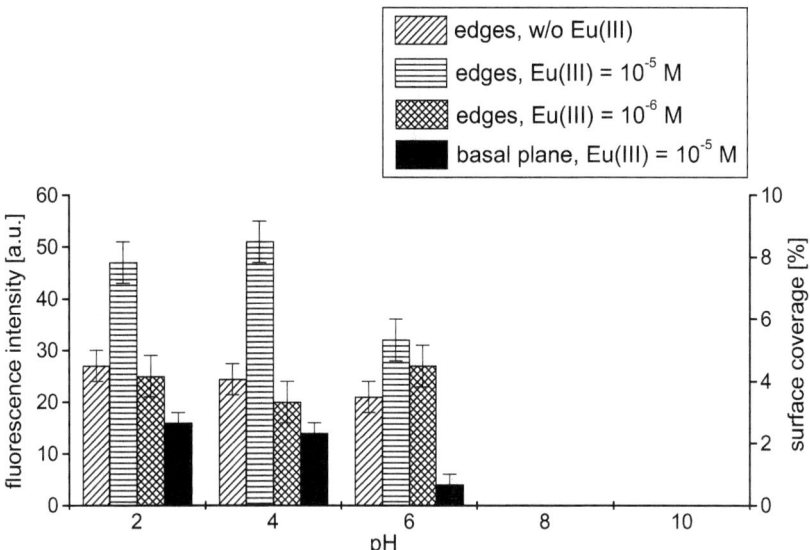

Fig. 24: Fluorescence intensity/surface coverage vs. pH for muscovite in the presence and absence of Eu(III): all measurements were undertaken at pH 2, 4, 6, 8, 10 (see text)

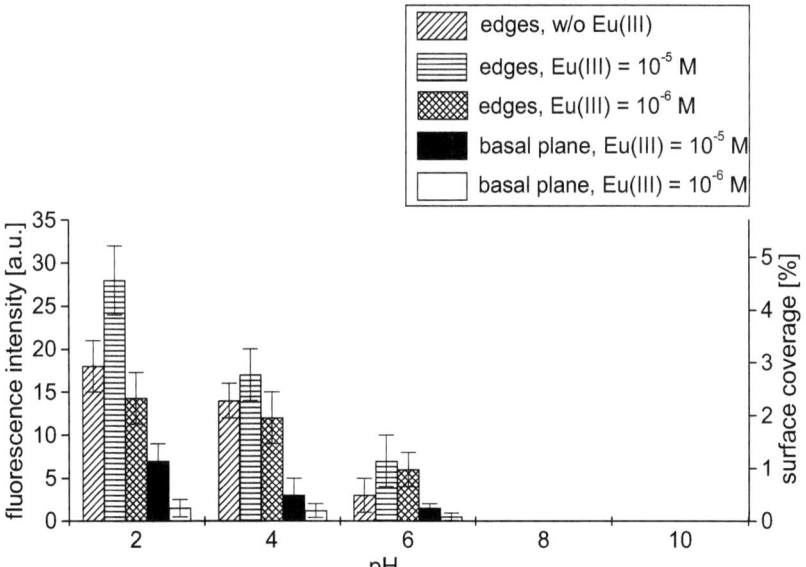

Fig. 25: Fluorescence intensity/surface coverage vs. pH of biotite in the presence and absence of Eu(III): all measurements were undertaken at pH 2, 4, 6, 8, 10 (see text)

Colloid adsorption in the absence of Eu(III): A significant fluorescence signal was detected on the mineral edges (diagonally lined bars in Fig. 24 and Fig. 25) under acidic conditions which decreased from pH 2 to 6. In contrast, under alkaline conditions, there was no fluorescence detected at the edges (pH 8-10, not shown). For details about the estimation of the detection limit see section IV-2.1.

Colloid adsorption in the presence of Eu(III): In the presence of Eu(III), a significant increase of fluorescence intensity was observed both on edges and planes. In Fig. 23 (b) it can be seen, that sorption also took place on the basal planes of muscovite. Black bars in Fig. 24 and Fig. 25 represent fluorescence intensities on basal planes in the presence of 10^{-5} M Eu(III). In the case of biotite, fluorescence was also detectable with a lower Eu(III) concentration (10^{-6} M) (white bars in Fig. 25).

At the edges of both sheet silicates increased fluorescence intensities were found in the presence of 10^{-5} M Eu(III) from pH 2 to 6 (horizontally lined bars in Fig. 24 and Fig. 25). At lower Eu(III) concentration (10^{-6} M) fluorescence intensities at the edges were for both minerals comparable to those obtained in absence of Eu(III) (cross-hatched bars in Fig. 24 and Fig. 25).

Colloid adsorption with Grimsel groundwater: No adsorption of colloids was detected on muscovite and biotite.

2.2.2 Feldspars (albite and K-feldspar)

The results of colloid adsorption experiments with albite and K-feldspar are depicted in Fig. 26 and Fig. 27. On albite fluorescence intensities/surface coverages did not differ significantly between mineral edges and surface planes. In contrast, on K-feldspar strong differences between edges and planes were observed.

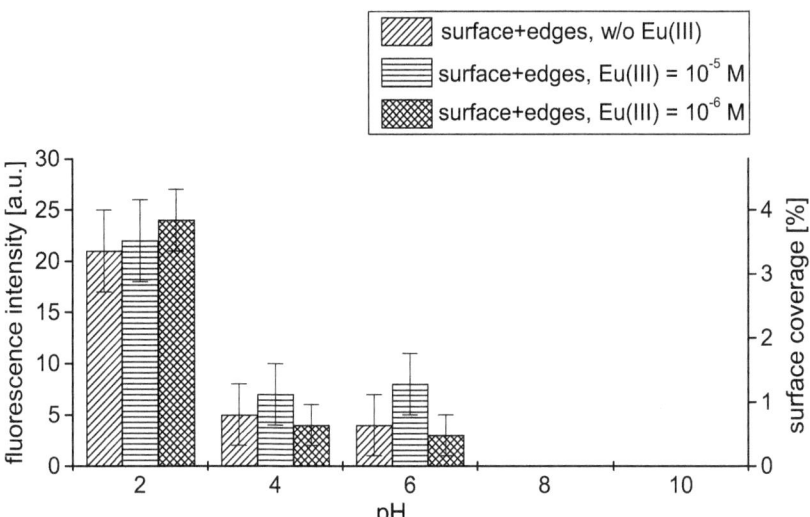

Fig. 26: Fluorescence intensity/surface coverage vs. pH for albite in the presence and absence of Eu(III): all measurements were undertaken at pH 2, 4, 6, 8, 10 (see text)

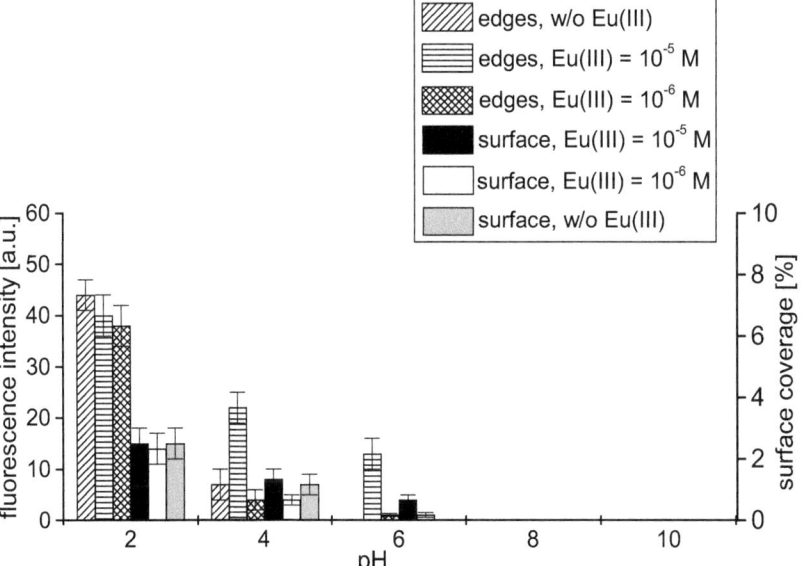

Fig. 27: Fluorescence intensity/surface coverage vs. pH of K-feldspar in the presence and absence of Eu(III): all measurements were undertaken at pH 2, 4, 6, 8, 10 (see text)

IV Results

Colloid adsorption in the absence of Eu(III): In the absence of Eu(III), fluorescence intensities on the edges as well as on the surface planes decreased with increasing pH (diagonally lined bar in Fig. 26 and diagonally/light grey bars in Fig. 27). In the alkaline regime no fluorescence was detected on the entire surface.

Colloid adsorption in the presence of Eu(III): The influence of Eu(III) strongly depended on pH. At pH 2 there was no significant influence of Eu(III). In the presence of 10^{-6} M Eu(III) observed fluorescence intensities were similar to those measured in the absence of Eu(III) (cross-hatched bar in Fig. 26 and additionally white bars in Fig. 27). Increasing the Eu(III) concentration to 10^{-5} M resulted in a slight increase in fluorescence intensity at pH 4 and 6 for albite (horizontal lined bars in Fig. 26. Under the same conditions K-feldspar showed increased fluorescence at pH 4 and pH 6, which was particularly apparent on the mineral edges (compare horizontally lined and diagonal bars in Fig. 27).

Colloid adsorption with Grimsel groundwater: No adsorption of colloids was detected on albite and K-feldspar.

2.2.3 Quartz

Colloid adsorption in the absence of Eu(III): Fig. 28 shows the fluorescence intensity/surface coverage vs. pH of the sorption experiment with quartz. In all experiments, the fluorescence intensities did not significantly differ between the surface planes and surface edges. In the absence of Eu(III), fluorescence was only detected at pH 2, a value close to the pH_{pzc} [52] (diagonally lined bar in Fig. 28).

Colloid adsorption in the presence of Eu(III): On addition of 10^{-6} M Eu(III) no significant effect was measured (cross-hatched bars in Fig. 28). By introduction of 10^{-5} M Eu(III), no significant effect on the measured fluorescence intensities was observed at pH 2. However, at pH 6, very weak fluorescence was measured (horizontally lined bars in Fig. 28).

Colloid adsorption with Grimsel groundwater: No fluorescence was detected.

IV Results 62

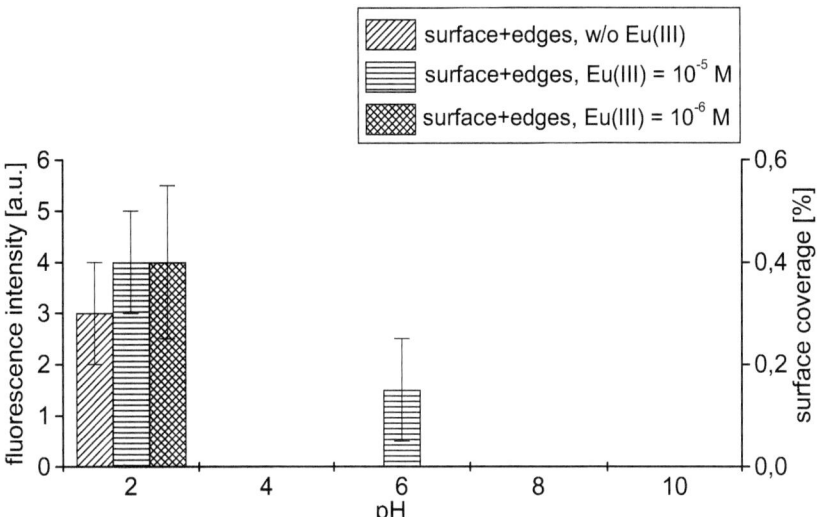

Fig. 28: Fluorescence intensity/surface coverage vs. pH of quartz in the presence and absence of Eu(III): all measurements were undertaken at pH 2, 4, 6, 8, 10 (see text)

2.2.4 Apatite

Colloid adsorption in the absence of Eu(III): Fig. 29 shows the results obtained for apatite. As in the case of albite, no significant differences of the fluorescence intensities/surface coverage were observed between mineral edges and surface planes. In the absence of Eu(III) (diagonally lined bars in Fig. 29), the highest fluorescence intensity was found at pH 2, comparatively lower fluorescence intensities were detected in the pH range from 4 to 8, a value close to the pH_{pzc} (about 7.6) [164]. At pH 10, no fluorescence was detected.

Colloid adsorption in the presence of Eu(III): Addition of 10^{-6} M Eu(III) had no significant effect on the measured fluorescence intensities (cross-hatched bar in Fig. 29). Increasing the concentration to 10^{-5} M Eu(III), a slight increase of fluorescence intensities was detected (horizontally lined bar in Fig. 29).

Colloid adsorption with Grimsel groundwater: Weak adsorption of colloids was detected on apatite as can be seen in Fig. 29.

IV Results

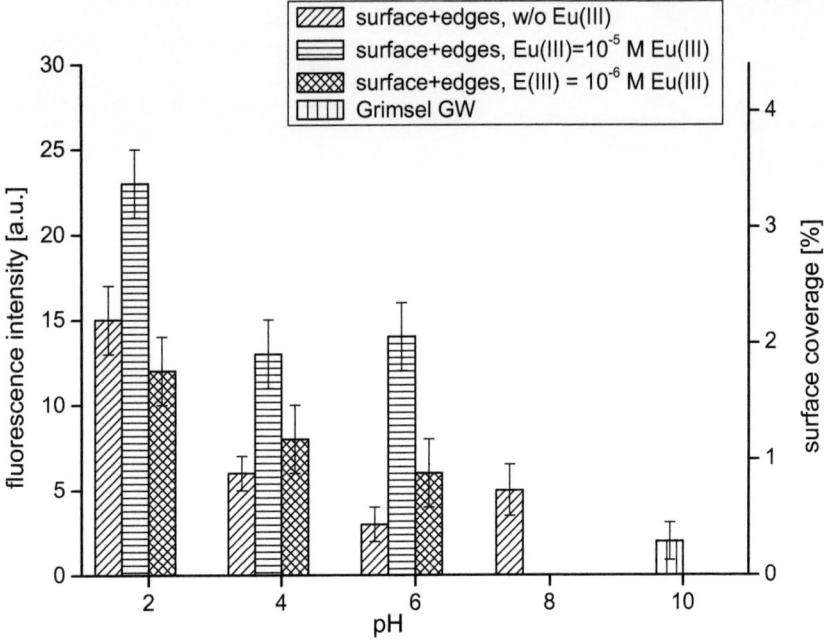

Fig. 29: Fluorescence intensity/surface coverage vs. pH of apatite in the presence and absence of Eu(III): all measurements were undertaken at pH 2, 4, 6, 8, 10 (see text)

2.2.5 Sapphire and titanite

Colloid adsorption in the absence of Eu(III): The results show that with background electrolyte only the fluorescence intensity on sapphire decreased with increasing pH (see Fig. 30, diagonally lined bars). No adsorption was detected at pH values ≥ 6. Also no fluorescence was detected in the experiments carried out with Grimsel groundwater.

Colloid adsorption in the presence of Eu(III): Experiments carried out with 10^{-5} and 10^{-6} M Eu(III) and sapphire show a significant influence on the measured fluorescence intensities at pH values ≥ 3.8. In experiments with the higher Eu(III) concentration the highest fluorescence intensities were measured (compare horizontally lined with cross-hatched bars in Fig. 30).

Colloid adsorption with Grimsel groundwater: No fluorescence was detected (not shown) on sapphire.

Additional colloid adsorption experiments were carried out with titanite and Grimsel groundwater. No fluorescence was detected.

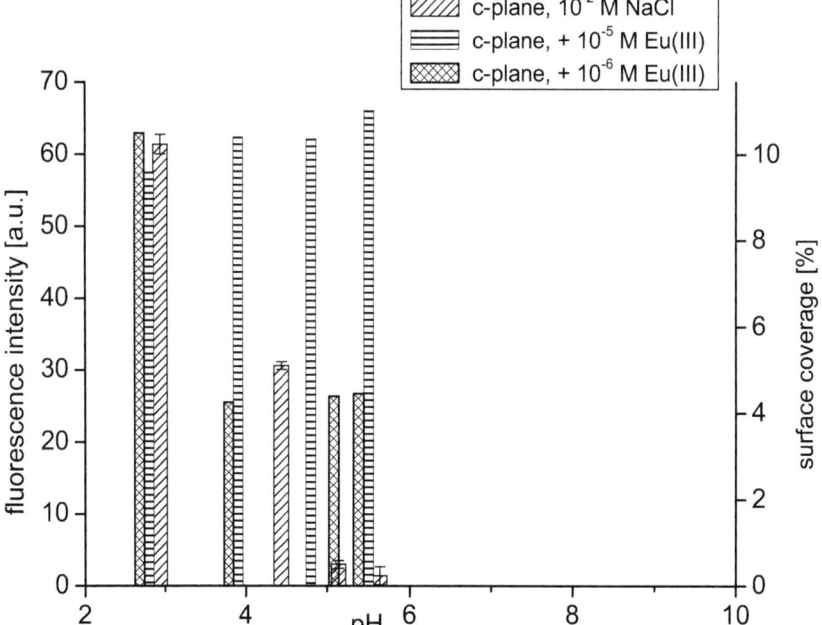

Fig. 30: Fluorescence intensity/surface coverage vs. pH of sapphire in the presence and absence of Eu(III) (see text)

2.3 Interaction of carboxylated latex colloids with Grimsel granodiorite

Based on the findings observed on the single minerals, similar sorption experiments were performed using Grimsel granodiorite and the fluorescent carboxylated latex colloids. The mineral samples were prepared from natural fracture filling material with a complex mineralogical composition (bulk composition see Table 1). Due to weathering processes, the mineralogical composition on the surfaces probably differs from the bulk composition, i.e. the content of weathering products like clay minerals will be increased. Also, the surface morphology will be more complex than on the single minerals including increased surface roughness, pores or cavities and impurities.

On these natural surfaces it was typically observed that increased fluorescence/colloid adsorption could be detected on surface areas with a pronounced morphology, e.g., holes, cavities or micro-fissures. As an example, Fig. 31 shows a section of the granodiorite surface characterized as albite by SEM/EDX (Fig. 31 (a)) and the corresponding fluorescence-optical image (Fig. 31 (b)).

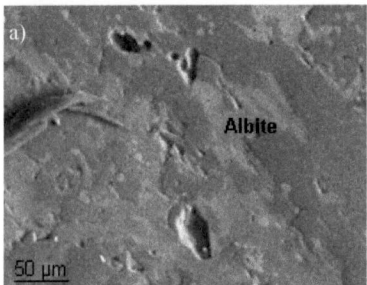 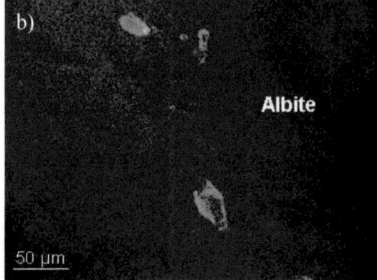

Fig. 31: (a) SEM image and (b) fluorescence-optical image of the granodiorite surface (pH = 4, I = 10^{-2} M NaCl, $c_{colloids}$ = 0.05 g/l), both at approximately the same position (see text)

Increased fluorescence intensities were found in cavities or holes (light areas in Fig. 31 (b)), whereas on the entire rock surface only weak fluorescence was detected under these conditions. As both plane surface areas and cavities provide the same mineralogy, one can conclude from this observation that sample morphology influences the process of colloid sorption. However, the difference in sorption processes between the entire surface and cavities is actually not well understood. It can only be speculated about an increased number of reactive surface sites, thin layers of alteration phases, different local surface charges, or even impurities on these sample sites.

One further general observation was that fluorescence was increased at the mineral edges whereas only weak or no fluorescence was detected on the entire mineral surface. Fig. 32 (a) and (b) show a SEM and a fluorescence-optical image of a Grimsel granodiorite surface site with the mineral biotite. Colloids were adsorbed in the presence of 10^{-5} M Eu(III). Increased fluorescence was observed at sample sites where edges were visible by SEM. This is in accordance with the experiments using single minerals. On the basal planes much weaker sorption was detected only in the presence of Eu(III).

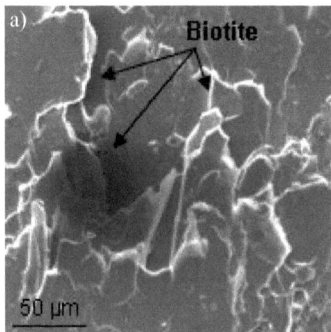 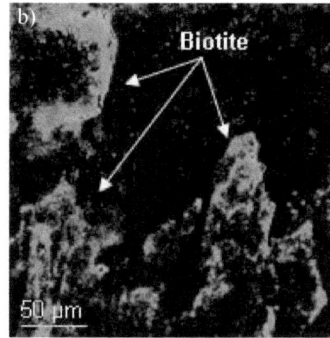

Fig. 32: (a) SEM-image and (b) fluorescence-optical image of the Grimsel granodiorite surface at approximately the same location. Sorption takes place preferably on the biotite edges, whereas fluorescence intensity on the planes is markedly weaker (pH = 4, I = 10^{-2} M NaCl, $c_{colloids}$ = 0.05 g/l , Eu(III) = 10^{-5} M) (see text)

In Fig. 33 (a) and (b), a light-optical and fluorescence-optical image of an apatite grain embedded in the granodiorite rock matrix is shown (pH 6, I = 10^{-2} M NaCl, $c_{colloids}$ = 0.05 g/l). This is an example of a distinct mineral phase where fluorescence was detected on the entire surface of an enclosed grain. Surrounding minerals were biotite and albite.

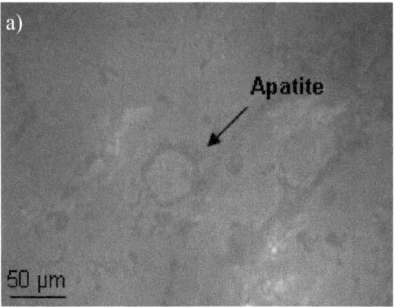 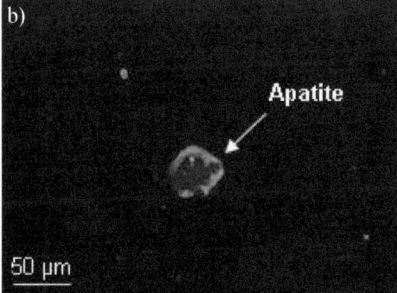

Fig. 33: (a) Light-optical image and (b) fluorescence-optical image of a granodiorite surface site; fluorescence was detected on the entire apatite surface (pH 6, I = 10^{-2} M NaCl, $c_{colloids}$ = 0.05 g/l) (see text)

The results of the sorption experiments with Grimsel granodiorite are summarized in Table 9.

Table 9: Adsorbing minerals and corresponding pH values on Grimsel granodiorite

pH	mainly adsorbing minerals (without Eu(III))	with 10^{-5} M Eu(III)
10 and 9.6 (Grimsel groundwater)	none	not measured
8	none	not measured
6	apatite, illite, titanite, biotite (edges)	whole surface
4	apatite, illite, titanite, K-feldspars, biotite (edges)	whole surface
2	apatite, illite, titanite, K-feldspars, biotite (edges)	whole surface

Fluorescence intensities on granodiorite surfaces were not quantified because, due to the complexity of the natural surface (e.g., roughness variations, mineral face geometry, surface weathering) fluorescence values strongly scattered for the mineral phases. From Table 9 only qualitative conclusions are possible.

The main adsorbing minerals were apatite, titanite, biotite (pH 2-6), K-feldspars (pH 2-4) and a phase, which is most likely illite. For apatite, titanite and K-feldspars, the EDX-data match well with the mineral sum formulas. For the illite-like phase, an elemental composition near the sum formula $K_{0,64}X_{0,10}(Al_{1,46}Fe^{3+}_{0,21}Fe^{2+}_{0,08}Mg_{0,28})(Al_{0,05}Si_{3,55})O_{10}(OH)_2$ given in Schachtschabel and Scheffer [51] was obtained, where X represents the exchangeable cations in equivalents. Due to an increased content of Mg and Fe found in the illite-like phase it is assumed that this phase could have been formed through weathering of biotite, which is a major bulk component of the granodiorite (Table 1).

Fluorescence was also detected on titanite, a rarely occurring mineral in the granodiorite. This mineral could not be found on the rock surfaces in the experiments under alkaline conditions. Introduction of 10^{-5} M Eu(III) showed colloid sorption on the whole rock surface in the investigated pH range of 2-6, whereas, in general, the highest fluorescence intensities were observed at pH 2. In all experiments, no significant influence of 10^{-6} M Eu(III) on the sorption behaviour was detected.

No fluorescence was detected at alkaline pH values, probably because of the dominating repulsive conditions (for details about the estimation of the detection limit see section IV-2.1.). This is in general accordance with the experiments on the single minerals.

2.3.1 Colloid desorption experiment

To gain insight into the colloid desorption kinetics and to ascertain whether the colloids are reversibly or irreversibly bound to the Grimsel granodiorite surface a colloid desorption experiment was carried out. The colloids were adsorbed at acidic pH and their desorption at alkaline pH was investigated (see section III-3.1 for experimental details).

In Fig. 34 the results of the colloid desorption experiment are shown. It can be seen that the fluorescence intensity measured on albite and biotite decreases rapidly in the first 10 days. After approximately 12 days only weak or no fluorescence could be detected. In general, after 25 days desorption time, no fluorescence could be detected on the whole rock surface.

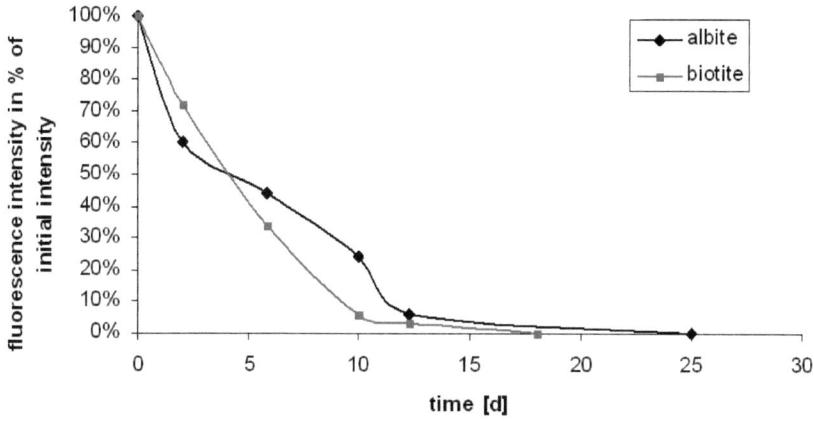

Fig. 34: Fluorescence intensity vs. time on the minerals albite and biotite (see text)

3. AFM force spectroscopy experiments

With the sorption experiments discussed above, valuable insight into colloid adsorption behaviour under various geochemical conditions was gained. However, from these investigations no mechanisms which could be responsible for the binding under unfavourable conditions (alkaline pH) could be identified. For this reason, complementary AFM force spectroscopy experiments were undertaken. With this method it was possible to identify and quantify attractive and repulsive forces resulting from the colloid-mineral surface interaction with high sensitivity. The force spectroscopy experiments were conducted with the colloid probe technique and different relevant mineral phases composing the Grimsel granodiorite (for experimental details see section III-3.3).

In the following section, the snap-in and adhesion forces derived from the force spectroscopy measurements will be presented separately. Additionally, approach curves measured at pH 10 on selected minerals in presence or absence of Ca(II) are compared with each other. Since Ca(II) is the dominating cation in Grimsel groundwater it is important to gain insight into the influence of this cation. As stated above, the experiments with UO_2^{2+} were carried out because Uranium is a main component of spent nuclear fuel rods and occurs under environmental conditions as mobile UO_2^{2+} ions.

The snap-in forces can be seen as the responsible forces for the colloid adsorption onto mineral surfaces. The adhesion force can be seen as the force required to detach an adsorbed particle from the mineral surface, for example by shear force due to groundwater flow. The discussion of the results can be found in section V. The error bars of the measured values are shown exemplary.

3.1 Sheet silicates (muscovite and biotite)

Fig. 35 and Fig. 36 show sample measurements on muscovite with Grimsel groundwater. Fig. 35 shows the raw data of the measurement (cantilever deflection vs. piezo position) and Fig. 36 shows the derived force-distance curve. The measurement shows that, although on approach no attractive forces were apparent, on retraction attractive adhesion forces could be detected.

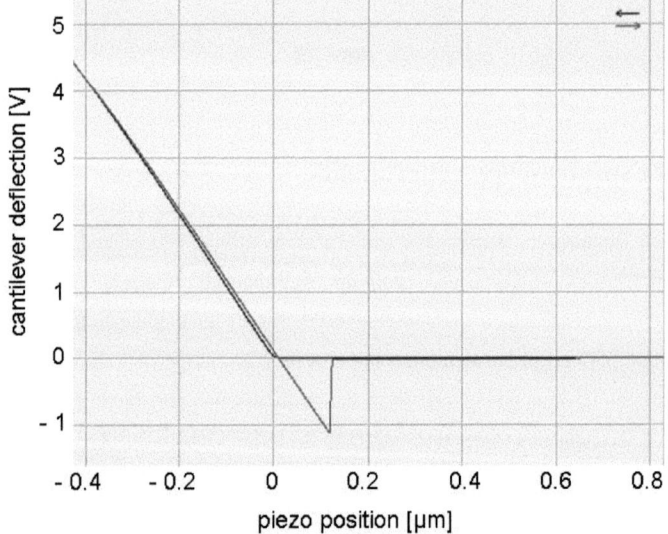

Fig. 35: Cantilever deflection vs. piezo position measured on muscovite with Grimsel groundwater (see text)

IV Results 70

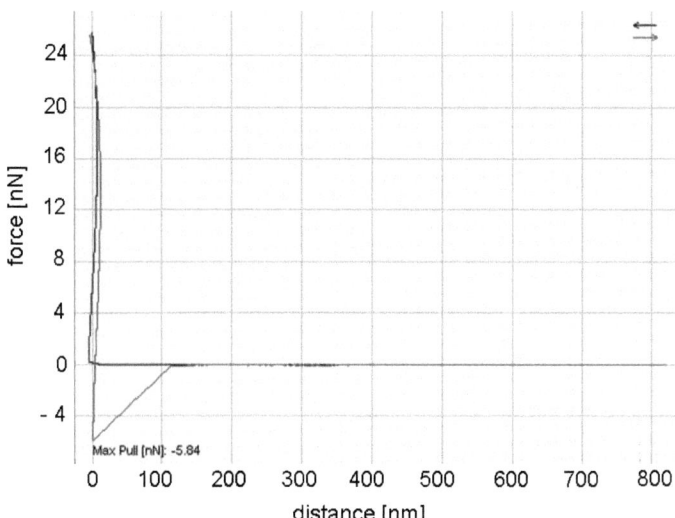

Fig. 36: Force-distance measurement (with baseline and hysteresis correction) measured on muscovite with Grimsel groundwater (see text)

Fig. 37 shows snap-in forces measured on muscovite. Black squares represent measurements in the presence of background electrolyte only. Snap-in forces were measured at pH 2 and 3. No obvious influence of Eu(III) and Ca(II) was observed at pH 2 and 3. In the presence of 10^{-5} M Eu(III) (grey diamonds) snap-in forces were also measured at pH 4-6 and were significantly increasing with pH. Experiments with 10^{-6} M Eu(III) (black triangles) showed only comparatively weak snap-in forces at pH 4-6. In the presence of 10^{-4} M Ca(II) (black asterisk), no snap-in forces were measured at pH values ≥ 4. Similarly, no snap-in forces were observed in the experiments carried out with 10^{-6} M UO_2^{2+} and Grimsel groundwater (not shown).

IV Results

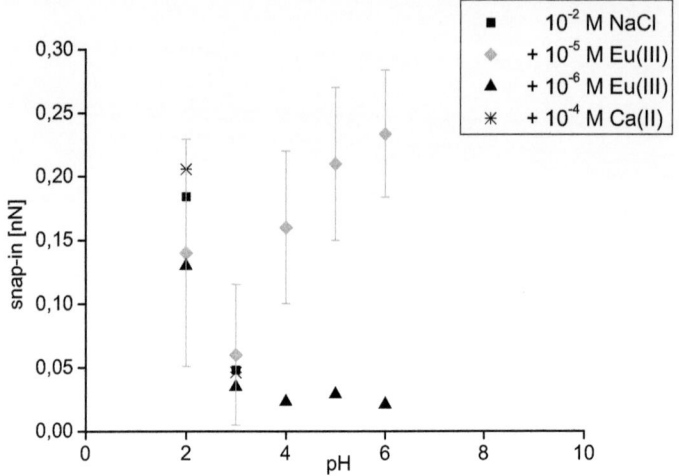

Fig. 37: Experimental snap-in forces on muscovite (see text)

Fig. 38 shows <u>adhesion forces</u> measured on muscovite. Measurements with background electrolyte showed decreasing adhesion forces with increasing pH (black squares). No attractive forces were detected at pH 8 and 10 (not shown). Again, no significant influence of cations was observed at pH 2 and 3. At pH 4-6 an increasing influence of Eu(III) was observed (grey diamonds/black triangles), whereas the higher Eu(III) concentration also had a significant influence on the measured adhesion forces. In the presence of 10^{-4} M Ca(II), adhesion forces at pH 4-10 were significantly increased (black asterisk). The same is true for the experiments with muscovite in presence of 10^{-6} M UO_2^{2+} at pH 4-6 (black diamonds). Experiments with Grimsel groundwater also showed increased adhesion forces (black cross) and were within the same range as the adhesion forces observed with Ca(II) at pH 10.

IV Results 72

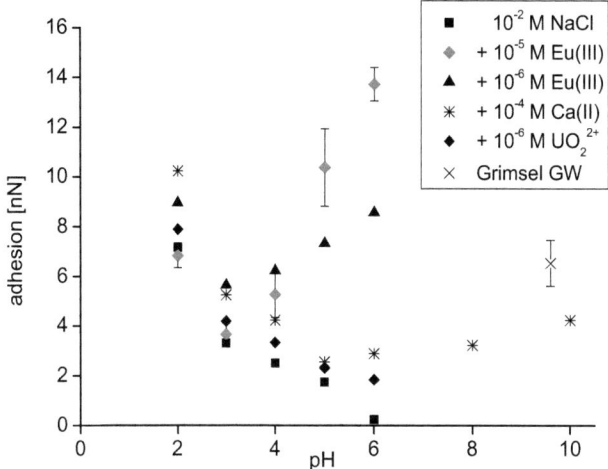

Fig. 38: Experimental adhesion forces on muscovite (see text)

Fig. 39 shows force-distance approach curves measured on muscovite. The red line represents the measurement at pH 10 and I = 10^{-2} M NaCl, the blue line represents the measurement under similar conditions but with 10^{-4} M Ca(II) added. It can be seen, that the magnitude of repulsive forces is clearly decreased in presence of Ca(II) (see section V).

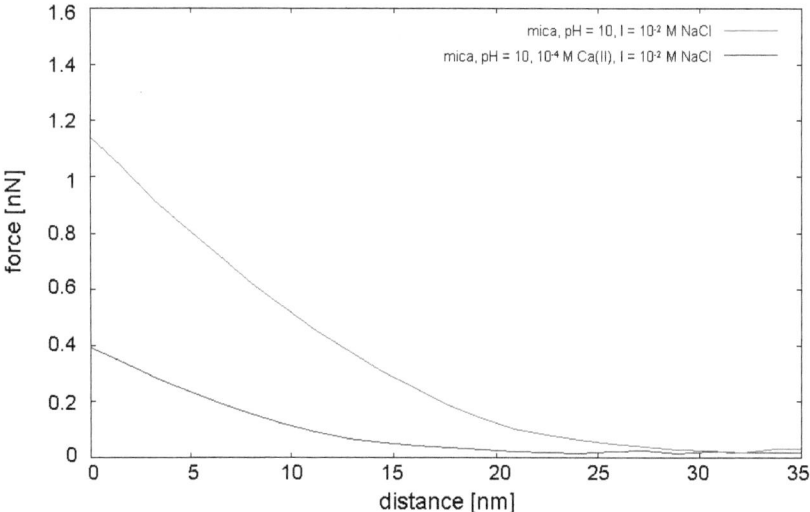

Fig. 39: Force-distance approach curves measured on muscovite (see text)

Fig. 40 shows measured snap-in forces on biotite. Black squares represent values of experiments carried out with background electrolyte only. Snap-in forces were observed at pH 2 and 3; repulsion sets in at pH 4-10. In the presence of 10^{-5} M Eu(III) (grey diamonds) no significant influence was observed at pH 2 and 3, whereas at pH > 4 increased snap-in forces were observed. The presence of 10^{-6} M Eu(III) also leads to a slight increase in snap-in forces at pH 4-6 (black triangles). In the presence of 10^{-4} M Ca(II) (black asterisk), 10^{-6} M UO_2^{2+} and with Grimsel groundwater no influence on the snap-in forces was observed (not shown).

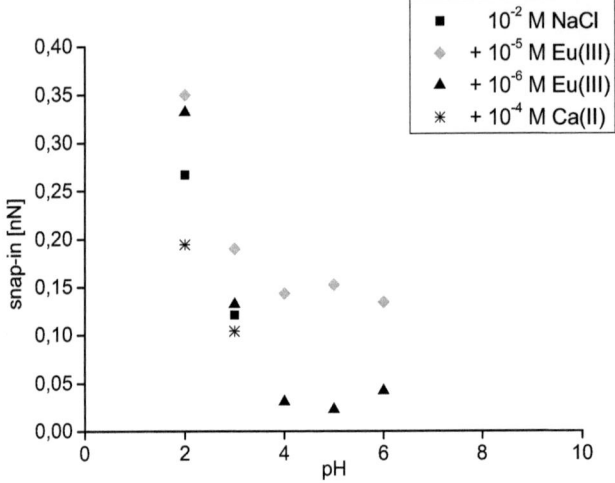

Fig. 40: Experimental snap-in forces on biotite (see text)

Fig. 41 shows measured adhesion forces on biotite. Black squares represent measured adhesion forces in the presence of background electrolyte only. Adhesion decreased with increasing pH and no attractive forces were measured at pH 8 and 10. In the presence of 10^{-5} M Eu(III) significantly increased adhesion forces were observed at pH 4-6 (grey diamonds). 10^{-6} M Eu(III) also increased adhesion forces at pH 4-6 (black triangles). 10^{-4} M Ca(II) increased the measured adhesion forces at pH 5-10 (black asterisk). With 10^{-6} M UO_2^{2+} an increase in adhesion forces was observed at pH 5 and 6. Significant adhesion forces were also observed with Grimsel groundwater (black cross) in about the same range as the measurements with 10^{-4} M Ca(II).

IV Results

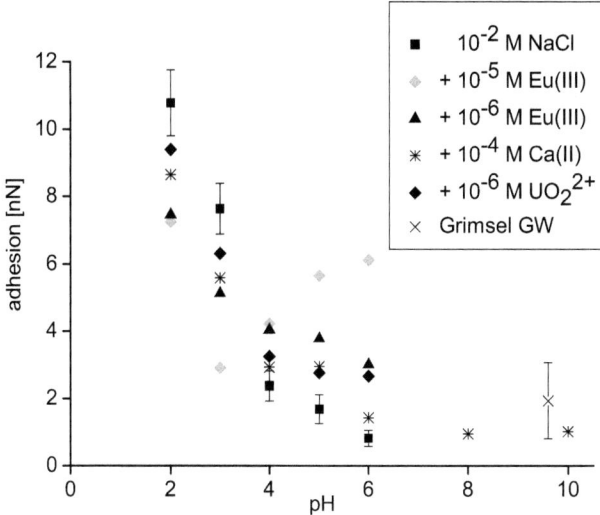

Fig. 41: Experimental adhesion forces on biotite (see text)

3.2 K-feldspar

Fig. 42 shows the measured snap-in forces measured on K-feldspar. Snap-in forces in the presence of background electrolyte (black squares) were measured at pH 2-4; at higher pH values > 4 only repulsion was observed (not shown). In the presence of 10^{-5} M Eu(III) (grey diamonds), a significant increase of the measured snap-in forces can be observed at pH 5 and 6. The lower Eu(III) concentration also increased the snap-in forces at pH 5 and 6, but to a comparatively lesser degree (compare grey diamonds/black triangles). In the presence of 10^{-4} M Ca(II) (black asterisk) no significant influence could be observed. Also with Grimsel groundwater no snap-in forces were detected.

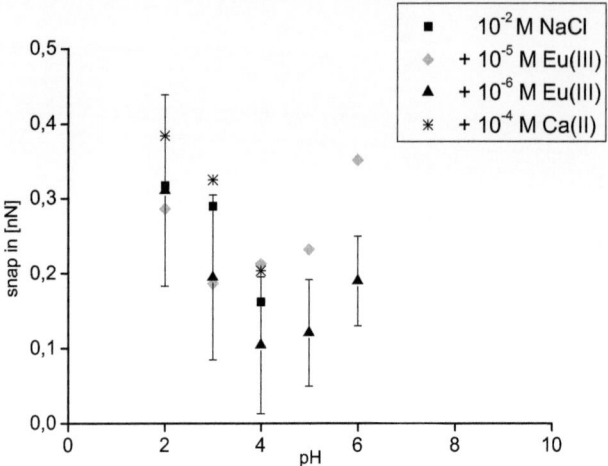

Fig. 42: Experimental snap-in forces on K-feldspar (see text)

Fig. 43 shows <u>adhesion forces</u> detected on K-feldspar. In the presence of background electrolyte, the adhesion forces decreased with increasing pH (black squares). At pH 10 adhesion forces were no longer evident. $10^{-5}/10^{-6}$ M Eu(III) had no significant influence at pH 2 and 3. At pH 4-6, an increasing influence of Eu(III) could be seen, whereas higher adhesion was observed at higher Eu(III) concentrations (compare grey diamonds/black triangles). In the presence of 10^{-4} M Ca(II) increased adhesion forces were observed from pH 4-10 (black asterisk). The experiment with Grimsel groundwater (black cross) showed similar adhesion values as the one carried out with 10^{-4} M Ca(II) at pH 10.

IV Results 76

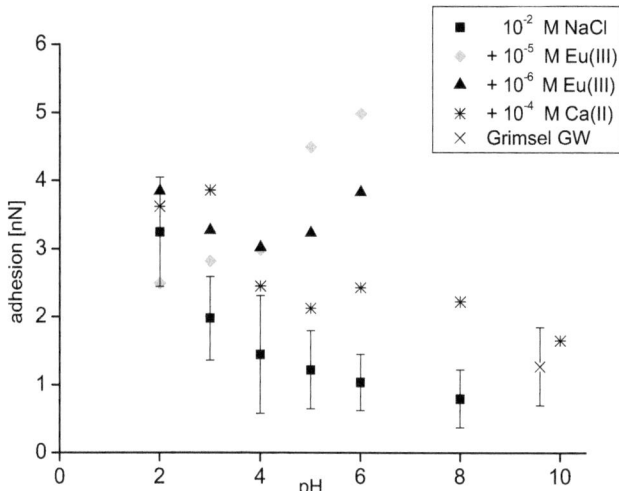

Fig. 43: Experimental adhesion forces on K-feldspar (see text)

Fig. 44 shows force-distance approach curves measured on K-feldspar. The red line represents the measurement at pH 10 and $I = 10^{-2}$ M NaCl, the blue line represents the measurement under similar conditions but with 10^{-4} M Ca(II) added. As in the measurements with muscovite, magnitude of repulsive forces is clearly decreased in presence of Ca(II) (see section V).

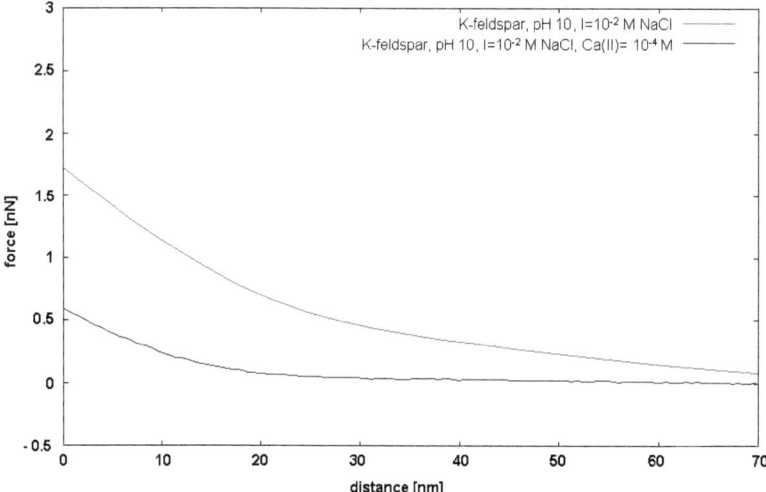

Fig. 44: Force-distance approach curves measured on K-feldspar (see text)

3.3 Quartz

Fig. 45 shows snap-in forces observed on quartz. In the presence of background electrolyte snap-in forces were only measured at pH 2 and 3 (black squares). At pH values ≥ 4 no attractive forces were measured. Ca(II) and Eu(III) obviously did not have a significant influence at pH 2 and 3 on the snap-in forces. $10^{-5}/10^{-6}$ M Eu(III) (grey diamonds/black triangles) increased the snap-in forces at pH 4-6, whereas an increasing trend can be observed with increasing pH. No obvious influence was observed in the presence of Ca(II) (black asterisk), UO_2^{2+} and with Grimsel groundwater (not shown).

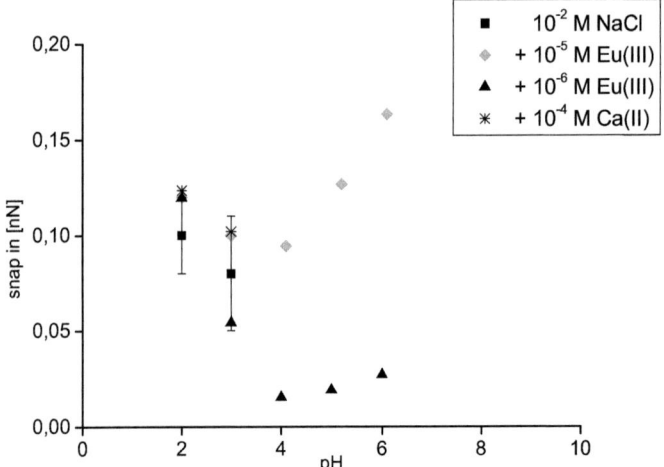

Fig. 45: Experimental snap-in forces on quartz (see text)

Fig. 46 shows the measured adhesion forces on quartz. In the presence of background electrolyte, maximum adhesion forces were measured at pH 2. These forces decrease with increasing pH and no adhesion was measured at pH 8 and 10. The presence of Eu(III) and Ca(II) did not have an obvious influence at pH 2 and 3. At pH 4-6, Eu(III) increased the adhesion significantly (grey diamonds/black triangles), whereas higher adhesion values were measured with the higher Eu(III) concentration. 10^{-4} M Ca(II) also increased the adhesion at pH 4-10 significantly (black asterisk). Similar observations were found in the measurements with 10^{-6} M UO_2^{2+}: adhesion was increased at pH values from 4-6. Experiments with Grimsel groundwater (black cross) also showed increased adhesion forces; again, the observed values were in the same range as those measured with Ca(II) at pH 10.

IV Results

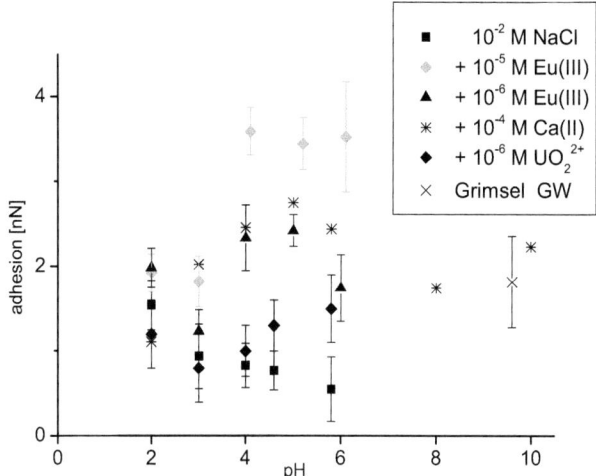

Fig. 46: Experimental adhesion forces on quartz (see text)

Fig. 47 shows force-distance approach curves measured on quartz. The red line represents the measurement at pH 10 and I = 10^{-2} M NaCl, the blue line represents the measurement under similar conditions but with 10^{-4} M Ca(II) added. As in the measurements on the minerals discussed above, the magnitude of repulsive forces is again clearly decreased in presence of Ca(II) (see section V).

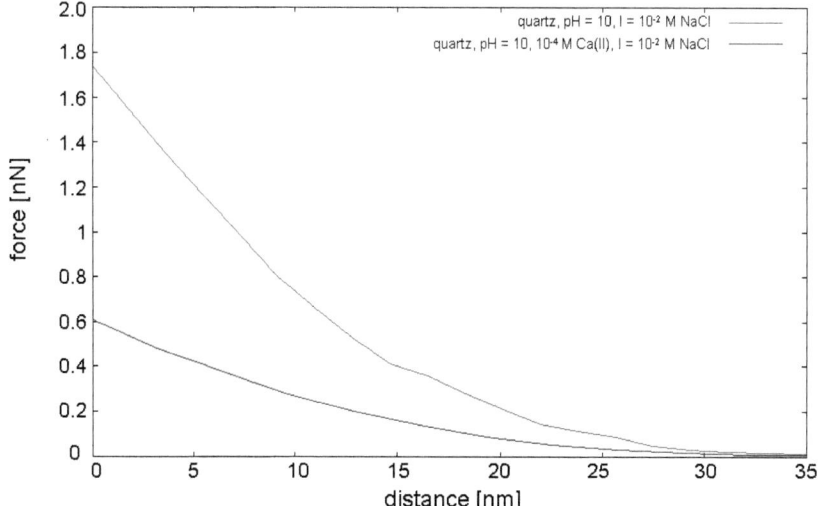

Fig. 47: Force-distance approach curves measured on quartz (see text)

3.4 Apatite

Fig. 48 shows the snap-in forces observed on apatite. In the range of pH 5 maximum snap-in forces were measured in the experiments with background electrolyte (black squares). Above and below this value, decreasing snap-in forces were observed. In presence of $10^{-5}/10^{-6}$ M Eu(III) no significant influence on the forces was seen (not shown). Experiments with 10^{-4} M Ca(II) also showed no significant influence at pH 2-8. However, at pH 10 attractive snap-in forces were detected (black asterisk). The experiments with Grimsel groundwater also showed snap-in forces in the same range as the experiments with 10^{-4} M Ca(II) at pH 10 (black cross). The experiments with 10^{-6} M UO_2^{2+} did not show any influence on the snap-in forces. In general, observed snap-in forces were highest for apatite as compared to the experiments with all other minerals.

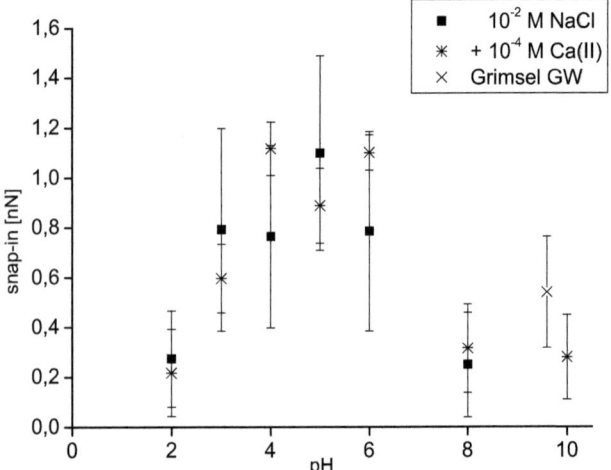

Fig. 48: Experimental snap-in forces on apatite (see text)

Fig. 49 shows the adhesion forces measured on apatite. Maximum values were observed at pH 5-6, minimal adhesion forces at pH 2. It can be seen that these forces also decreased with increasing and decreasing pH. Experiments carried out in the presence of $10^{-6}/10^{-5}$ M Eu(III) did not have any obvious influence on the adhesion forces (not shown). In presence of 10^{-4} M Ca(II) there was also no obvious influence observed at pH 2-8. Only at pH 10 an increase of adhesion forces was seen (black asterisk) compared to those measurements in the presence of background electrolyte. The experiments with 10^{-6} M UO_2^{2+} did not show any influence on the adhesion forces. Comparatively increased adhesion forces were also measured with Grimsel groundwater; adhesion forces were in the same range as the measurements with 10^{-4} M Ca(II) at pH 10 (black cross).

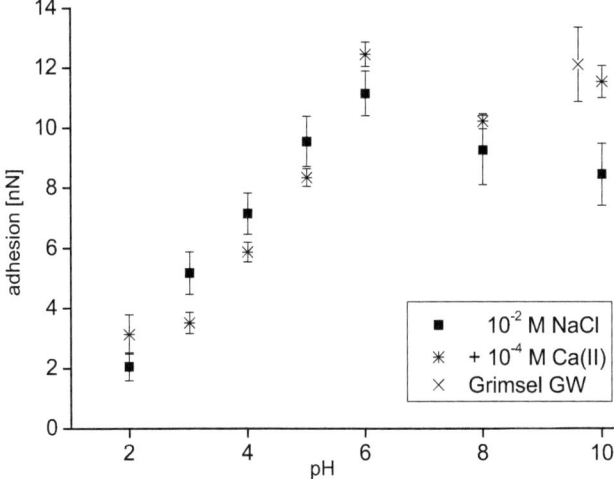

Fig. 49: Experimental adhesion forces on apatite (see text)

To gain insight into the dissolution behaviour of apatite, additional AFM contact mode images were recorded on several surface areas (each of 25 µm²) at pH 6, 4 and 2 with 10^{-2} M NaCl background electrolyte. Fig. 50 (a) and (b) show AFM height/deflection images measured at pH 6 after 10 min. immersion in solution. The surface was still intact and the measured RMS roughness of about 2.5 nm was in the range of value measured in air (see Table 7). Atomic steps are visible with a minimum step height of around 3 Å. The same area scanned at pH 4 (after 5 min under solution) already showed increased RMS roughness values of 4.1-5.3 nm (not shown). Fig. 51 (a) and (b) show AFM height/deflection images measured at pH 2 (immersion time ca. 1 min). It can be clearly seen that the surface was extensively corroded. The RMS roughness was significantly increased with values of about 15 nm (see section V).

IV Results

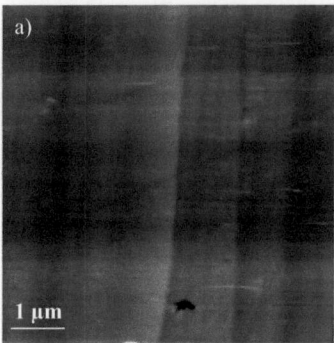 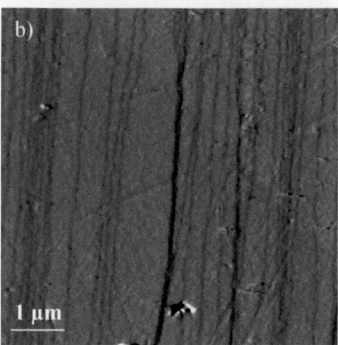

Fig. 50 (a) and (b): Height and deflection image of apatite after 10 min immersion in a solution of 10^{-2} M NaCl, pH 6 (see text)

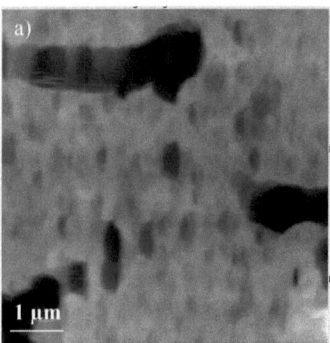 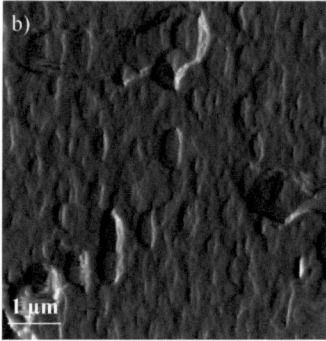

Fig. 51 (a) and (b): Height and deflection image of apatite after 1 min immersion in a solution of 10^{-2} M NaCl, pH 2 (see text)

3.5 Titanite

Additional experiments were conducted with titanite and Grimsel groundwater. No attractive snap-in forces were measured under all experimental conditions. Mean adhesion forces were around 2.7 nN ± 0.5 nN. Fig. 52 and Fig. 53 show force-distance raw data (cantilever deflection vs. piezo position) and its corresponding force-distance curve measured on titanite with Grimsel groundwater.

IV Results 82

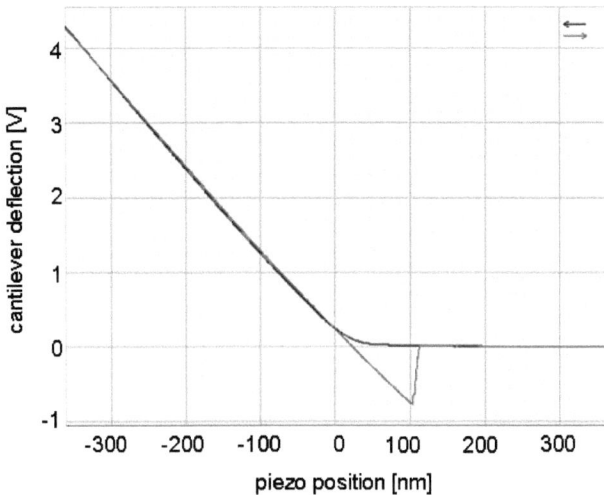

Fig. 52: Cantilever deflection vs. piezo position measured on titanite (Grimsel groundwater) (see text)

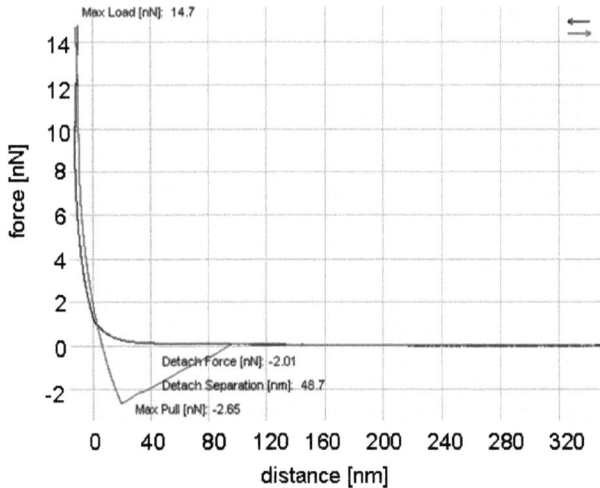

Fig. 53: Force-distance curve (with baseline and hysteresis correction) measured on titanite (Grimsel groundwater) (see text)

3.6 Force-volume measurements

By using AFM force-volume mode force-curves can be acquired on a whole sample area (for details see section III-1.6.3). Force-volume measurements with several mineral surfaces were conducted in order to identify possibly present surface charge heterogeneities. For the force-volume experiments the pH was set to 5.8, ionic strength is 10^{-2} M NaCl and the Ca(II) concentration was set to 10^{-4} M. As a representative example for these measurements, Fig. 54 shows an adhesion force map measured on quartz. The maximum range of the pull-off forces is about 1 nN (minimum values are about 2.5 nN; maximum values about 3.5 nN). No snap-in forces were detected on the whole area. The experiments with the other mineral samples also showed no significant deviation from the results presented in sections IV-3.1-3.5. Table 10 shows a summary of the observed snap-in and adhesion forces on the different mineral surfaces.

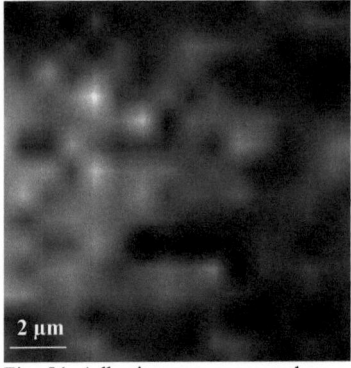

Fig. 54: Adhesion map measured on quartz. Values range from ca. 2.5 (dark areas) to 3.5 nN (bright areas) (see text)

Table 10: Results of the force-volume measurements; average values of 256 measurements per area

Mineral	snap-in [nN]	σ	adhesion[nN]	σ
quartz	0		2.88	0.19
biotite	0		2.53	0.3
muscovite	0		2.13	0.28
K-feldspar	0		2.76	0.96
apatite	1.21	0.37	8.47	1.42

4. Calculation of DLVO interaction force profiles

4.1 Contact angle measurements, surface energies and Hamaker constants

Contact angle measurements are needed to determine the surface energies of the individual mineral samples as outlined in section III-1.3. For this purpose the Hamaker constants of the individual samples were calculated by the method of van Oss [108] as described in sections II-2.6.2 and II-2.6.3. Table 11 shows the results of these measurements.

Table 11: Contact angles [°] measured on the different mineral surfaces with three solvents of varying polarity

mineral	water, pH 2 & 6	dimethyl sulphoxide	toluene
quartz	30 ± 3	15 ± 3	12 ± 2
muscovite	4	4	18
biotite	4	4	14
apatite	55	15	13

Contact angles measured on carboxylated polystyrene were taken from the literature; the following contact angles were measured: with water 87.6° ± 1.7°, with formamide 71.9° ± 1.4° and with diiodomethane 27.9° ± 1.4° [165].

By inserting the contact angles in the Young-Dupré equation (equation 20), the surface tension components could be calculated, as shown in Table 12. By inserting then the calculated Lifshitz-van der Waals surface tension components γ^{LW} in equation 27, the individual Hamaker constants were calculated. As is shown in Table 12, the literature values of the individual Hamaker constants are close to the calculated ones. Note that the Hamaker constant is not a sensitive parameter for the DLVO calculations and potential determination since it only has influence on the depth of the primary minimum. The Hamaker constant of water was taken from the literature (3.7×10^{-20} J, [91]).

IV Results

Table 12: Calculated surface tension components and Hamaker constants of the mineral samples

material	γ^{LW} [mJ/m²]	γ^+ [mJ/m²]	γ^- [mJ/m²]	surface energy, [mJ/m²]	calculated Hamaker constant [J]	Hamaker constant literature value [J]
quartz	34.4	3.65	48.76	70.98	$6.6*10^{-20}$	$5\text{-}6.5*10^{-20*}$
muscovite	34.6	5.07	125	85.1	$6.7*10^{-20}$	$10*10^{-20*}$
biotite	36	5.86	127.24	90.62	$6.9*10^{-20}$	?
apatite	27.07	0.05	38.9	29.79	$5.2*10^{-20}$	?
carboxylated polystyrene	45.07	1.26	5.88	50.52	$8.7*10^{-20}$	$6.6\text{-}7.9*10^{-20*}$

*[183]

By inserting the determined Hamaker constants into equation 13, the Hamaker constants for the individual mineral-water-carboxylated polystyrene colloid system could be calculated. The results of these calculations are shown in Table 13. In the literature only one value for the Hamaker constant of a mineral-water-colloid system, that of quartz, could be found.

Table 13: Calculated Hamaker constants for the mineral-water-colloid systems

mineral	calculated Hamaker constant [J]	Hamaker constant, literature value [J]
quartz	$6.6*10^{-21}$	$3.8*10^{-21*}$
muscovite	$6.8*10^{-21}$	?
biotite	$7.14*10^{-21}$	?
apatite	$3.6*10^{-21}$	?

*[166]

4.2 DLVO calculations

The total interaction force between the polystyrene microsphere attached to the AFM cantilevers and the mineral surfaces was calculated as the sum of forces describing electric double layer (F_{EL}), van der Waals interactions (F_{vdW}) and Born repulsion (F_{Born}):

$$F_{Total} = F_{El} + F_{vdW} + F_{Born} \quad (41)$$

Equations for a force were obtained from the corresponding equations for a energy of interaction between a sphere and a flat plane (see equations 6, 14 and 18) [91]:

$$F = -\frac{d\Delta G}{dx} \quad (42)$$

The electrical double layer interaction force was calculated by inserting equation 6 into equation 42:

$$F_{el}(x) = \kappa 64\pi\varepsilon\varepsilon_0 R\left(\frac{K_B T}{ze}\right)^2 \tanh\left(\frac{ze\Psi_1}{4K_B T}\right)\tanh\left(\frac{ze\Psi_2}{4K_B T}\right)\exp(-\kappa x) \quad (43)$$

The van der Waals interaction force was calculated by inserting equation 14 to equation 42:

$$F_{vdw}(x) = \frac{AR}{6x^2(1+(14x/\lambda))^2} + \frac{AR(14/\lambda)}{6x(1+(14x/\lambda))^2} \quad (44)$$

The Born interaction force was calculated by inserting equation 18 into equation 42:

$$F_{Born}(x) = -\frac{A\sigma^6}{7560}\left[\frac{1}{(2R+7)^7} - \frac{7(6R-x)}{x^8} - \frac{1}{x^7}\right] \quad (45)$$

4.2.1 Theoretical and experimental AFM force-distance curves

A comparison of theoretically and experimentally measured force-distance approach curves was undertaken to investigate the mineral surface potential and to find out if there are also non-DLVO forces (e.g., hydration forces) present under the different geochemical conditions. The surface potentials of selected minerals were elucidated by fitting the results of DLVO calculations to the experimental AFM force spectroscopy approach curves (equation 41 shows of which forces the DLVO approach curves are composed). Note that this procedure does not work for the retraction curves, since the colloid-collector interaction after establishing contact can not be modelled by DLVO theory.

DLVO fits to the experimental data were obtained by allowing the surface potentials of the minerals to vary with experimental conditions, while the surface potential of the individual colloid surface was assumed to be equal to the measured zeta potential. Note that due to measurement restrictions the resolution (i.e., the distance between sample points of the force curve) of the experimental force curves was not always optimal: the true resolution of a force measurement can initially be seen after a time consuming conversion procedure of the raw data.

As stated in section III-1.1.2 another way to determine surface potentials is to perform streaming potential measurements with the individual minerals. Both, the experimentally and theoretically determined surface potentials can be compared to derive conclusions about the nature of the encountered surface forces.

In the following section, selected force measurement data are compared to calculated force curves for different single minerals.

In all calculations the zeta potential of the polystyrene colloids measured for the respective experimental conditions was taken as the surface potential of the colloids. The surface potentials for the mineral surfaces were fitted to the experimental force-distance curves. Where available, those mineral surface potentials were compared to measured streaming potentials (in the case of quartz) or literature data.

Quartz

Fig. 55 shows experimental vs. calculated and fitted force-distance approach curves (black and red lines) on quartz. Diagonal crosses represent measurements undertaken at pH 6, $I = 10^{-2}$ M NaCl. Blue crosses represent measurements at pH 3 with similar ionic strength.

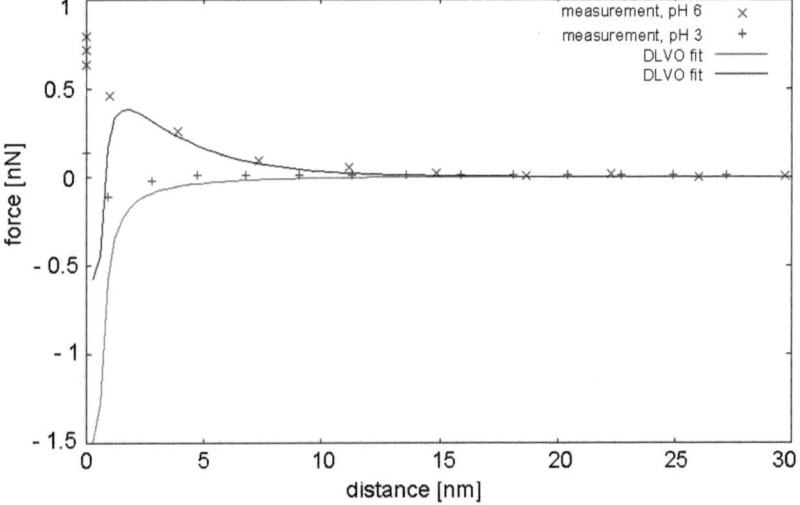

Fig. 55: Comparison of experimental approach data (pH 3 and pH 6, $I = 10^{-2}$ M NaCl) measured on quartz with DLVO calculations: the red line represents the DLVO fit to the measurement at pH 3, the black line represents the DLVO fit to the measurement at pH 6 (see text)

Fig. 56 shows experimental vs. calculated and fitted force-distance approach curves on quartz (black and red lines). Diagonal crosses represent measurements undertaken at pH 6, Eu(III) = 10^{-5} M, $I = 10^{-2}$ M NaCl. Blue crosses represent measurements with Grimsel groundwater.

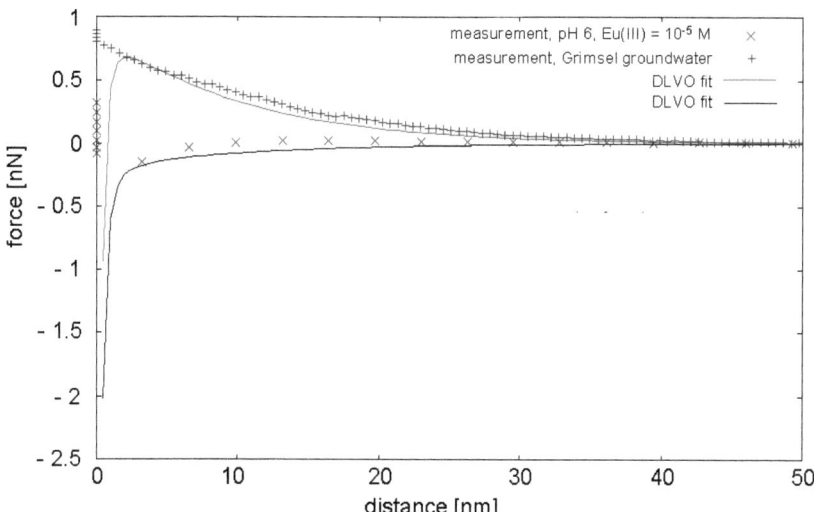

Fig. 56: Comparison of experimental approach data (pH = 6, 10^{-5} M Eu(III), I = 10^{-2} M NaCl and Grimsel groundwater) measured on quartz surface with DLVO calculations: the red line represents the DLVO fit to the measurements with Grimsel groundwater, the black line represents the DLVO fit to the measurement at pH 6, 10^{-5} M Eu(III) (see text)

Muscovite

Fig. 57 shows experimental vs. calculated and fitted force-distance approach curves on muscovite (black and red lines). Diagonal crosses represent measurements undertaken at pH 6, I = 10^{-2} M NaCl. Blue crosses represent measurements at pH 2 with similar ionic strength.

IV Results

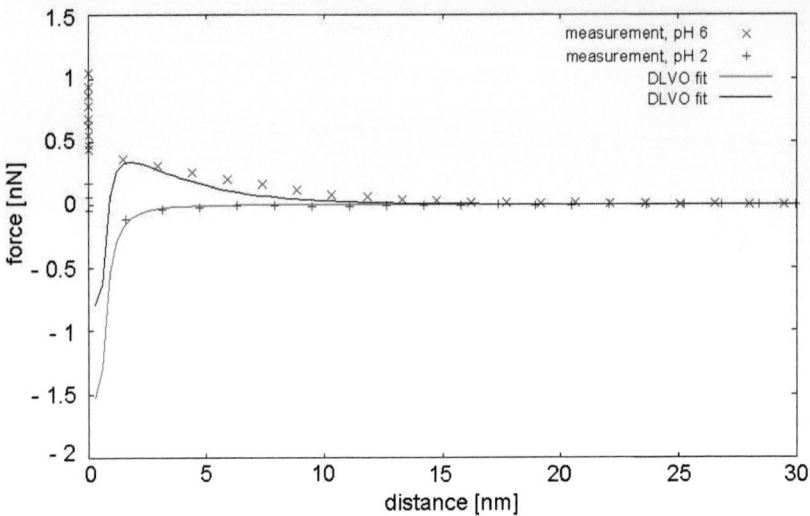

Fig. 57: Comparison of experimental approach data on muscovite surface with DLVO calculations (pH 2/6, I = 10^{-2} M NaCl): the red line represents the DLVO fit to the measurement at pH 2, the black line represents the DLVO fit to the measurement at pH 6 (see text)

Fig. 58 shows experimental vs. calculated and fitted force-distance approach curves on muscovite (black and red lines). Diagonal crosses represent measurements undertaken at pH 6, Eu(III) = 10^{-5} M, I = 10^{-2} M NaCl. Blue crosses represent measurements with Grimsel groundwater.

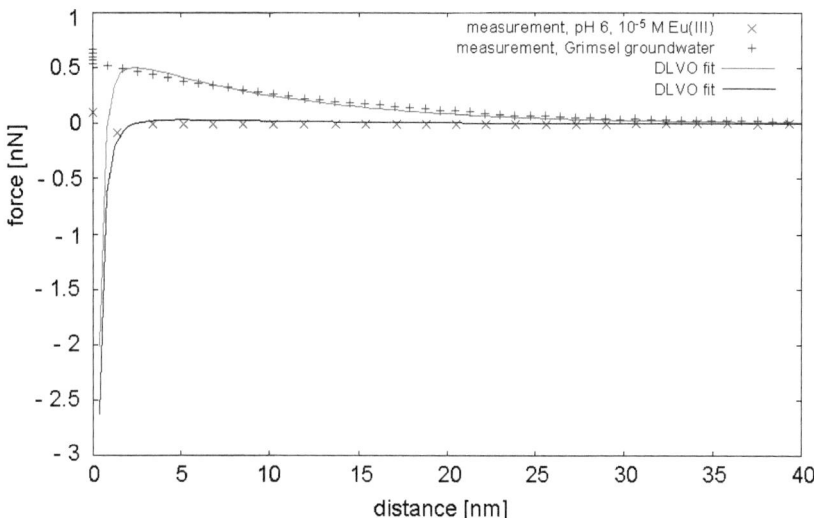

Fig. 58: Comparison of experimental approach data on muscovite surface with DLVO calculations (pH 6, I = 10^{-2} M NaCl and Grimsel groundwater): the red line represents the DLVO fit to the measurement with Grimsel groundwater, the black line represents the DLVO fit to the measurement at pH 6 (see text)

Biotite

Fig. 59 shows experimental vs. calculated and fitted force-distance approach curves on biotite (red lines). Blue crosses represent measurements with Grimsel groundwater.

IV Results

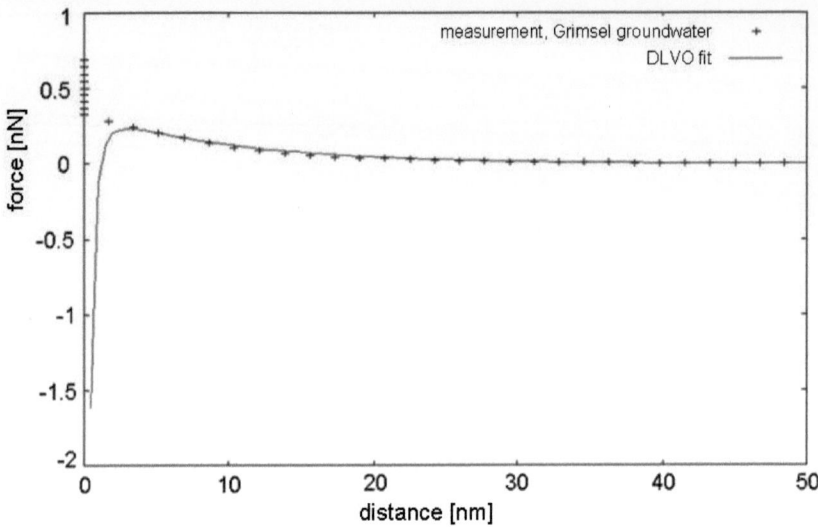

Fig. 59: Comparison of experimental approach data on biotite surface with DLVO calculations (Grimsel groundwater) (see text)

Apatite

Fig. 60 shows experimental vs. calculated and fitted force-distance approach curves on apatite (red lines). Blue crosses represent measurements with Grimsel groundwater.

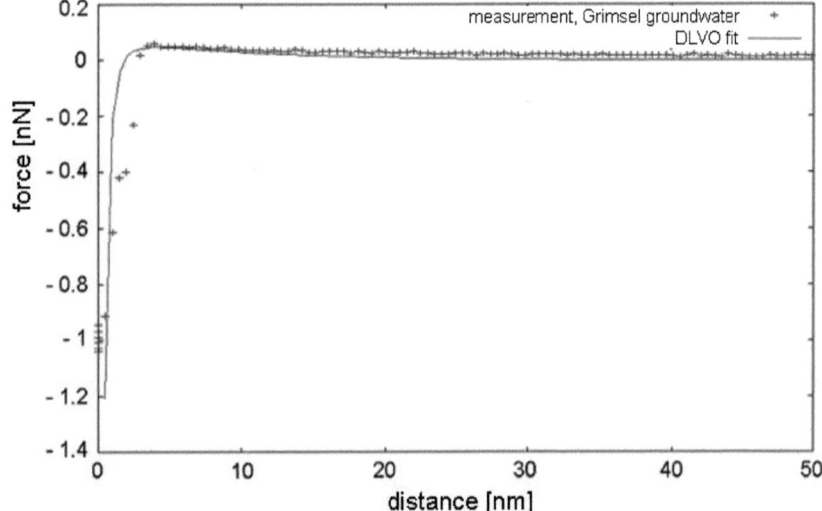

Fig. 60: Comparison of experimental approach data on apatite surface with DLVO calculations (Grimsel groundwater) (see text)

For all minerals under consideration, the experimental force distance curves could be reasonably described by the DLVO calculations except at distances < 5 nm. While DLVO theory suggests the existence of a deep primary minimum due to van-der-Waals attraction, such deep minima could not be experimentally established (except for apatite). For muscovite and quartz, attractive forces were found at pH 2 (and 3, respectively) and overall repulsive forces at pH 6 and 10 (see also Fig. 39 and Fig. 47). In the presence of 10^{-5} mol/L Eu(III) (pH 6), both measured and calculated force distance curves revealed an attractive regime at short distance. No such attraction could be observed in Grimsel groundwater even though the repulsive forces decreased as compared to a pure 10^{-2} M NaCl solution at pH 10 (compare Fig. 56 and 58 with Figs. 47 and 39).
In experiments with quartz, the fitted surface potentials showed good agreement with measured streaming potentials. Data for muscovite resemble quite well those found for quartz. Only the force-distance data obtained for apatite show attractive forces in Grimsel groundwater after overcoming a weak repulsive regime with maximal repulsive forces at around 5 nm. The calculated surface potential under those conditions lies around 0 +/- 2 mV. All fitted and measured surface potentials obtained in the present study are collected in Table 14.

Table 14: Fitted and measured surface potentials; values in brackets correspond to measured zeta- and streaming potential data

surface potentials [mV]	latex colloid	quartz	muscovite	biotite	apatite
pH = 3; I = 10^{-2} M NaCl	(-41)	+18 (+20)	+10		
pH = 6; I = 10^{-2} M NaCl	(-62)	-44 (-50)	-30		
pH = 6; I = 10^{-2} M NaCl; 10^{-5} M Eu(III)	(-30)	+10 (+20)	+3		
Grimsel groundwater; pH = 9.6	(-67)	-50	-40	-30	0

4.2.2 Maximum potential energy barrier calculations

Maximum potential energy barrier calculations were carried out in order to make a statement about the probability of colloid attachment under generally repulsive (alkaline) conditions. The calculations were performed with the theoretical (fitted) DLVO approach curves of the measurements with Grimsel groundwater and quartz, biotite, muscovite and apatite, respectively. Note that the determination of maximum potential energy barrier values cannot be undertaken for approach curves showing only repulsive interaction because in this case the maximum repulsive barrier would be infinitely high. In curves showing only attractive interaction, the energy barrier value is zero. Therefore, no calculations were carried out with the corresponding experimental force curves. The results of the calculations are summarized in Table 15. It can be seen that the highest energy barrier (and thus the smallest colloid attachment probability) exists for quartz under the measurement conditions. Since apatite has the smallest potential energy barrier, the highest colloid attachment probability can be expected on this mineral.

Table 15: Calculated maximum energy barriers for selected force spectroscopy measurements

Grimsel groundwater	maximum potential energy barrier DLVO (Φ/kT) [ ]
muscovite	1058
biotite	782
quartz	1912
apatite	160

V Discussion

In the present work a novel approach for a better understanding of colloid adsorption processes on mineral surfaces was developed and successfully applied. In principal this approach consists in the combination of fluorescence optical, scanning electron and atomic force microscopy and spectroscopy with adsorption theory.

Adsorption of fluorescent carboxylated latex colloids on Grimsel granodiorite and a number of its single component minerals was studied by fluorescence microscopy and SEM/EDX. Although single colloids (< 500 nm) can not be resolved with the experimental setup, it was demonstrated to be very sensitive to quantify colloid adsorption processes. As a complementary technique, AFM force spectroscopy with carboxylated latex colloids – used as colloid probes - was used. With this method it was possible to quantify and identify forces responsible for the colloid interaction with mineral surfaces with high sensitivity. Sorption and force spectroscopy experiments were conducted under variation of the geochemical conditions.

The latex colloids were chosen as model for natural colloids (e.g., bentonite colloids) since both are negatively charged over a wide pH range and the differences of their surface potentials are not too high (see Fig. 19 and Fig. 20). However, one has to make a distinction regarding the comparability between natural and synthetic colloids: the spherical latex colloids do not exhibit the heterogeneous charge distribution typical of natural clay colloids with their differing charges on edges and planes.

The zeta-potentials of the latex colloids used in the sorption experiments differed significantly from those used in the force spectroscopy experiments. One reason for this is possibly the differing carboxyl group density on the colloids.

It has to be noted, that colloid attachment and agglomeration are kinetically controlled phenomena. Since there is always a certain "sticking" probability that colloids may overcome a repulsive energy barrier in dependence of time and hence adsorb onto the surfaces, the short-term experiments carried out in this work are not transferable onto long time scales.

As stated above, adhesion and snap-in forces were directly quantified by AFM force spectroscopy. As a reminder, the adhesion force can be seen as the force required to detach an adsorbed particle from the mineral surface, for example by shear force due to groundwater

V Discussion

flow. The snap-in forces can be seen as the responsible (van der Waals) forces for the colloid adsorption onto mineral surfaces. The adhesion forces are always greater than the snap-in forces. This occurs for several reasons: (a) during the colloid probe contact with the surface, some chemical bonds (e.g., hydrogen bonds, covalent bonds) or adhesive bonds may elicit non-DLVO forces. (b) during the contact, the colloid probe and/or the sample may be deformed, increasing the contact area, which can lead to increased adhesion forces. Furthermore, it is known that adhesion forces decrease with increasing surface roughness (= decreasing contact area) [183]. It can also be assumed that surface roughness may exhibit heterogeneously distributed surface charges which may increase or decrease the snap-in forces as compared to measurements undertaken on smooth surfaces. Note that not only mineral surfaces but also the carboxylated colloids exhibit a certain roughness. Roughness measurements by Tormoen and Drelich of a 14 µm latex bead shows asperities in the order of several to a few tenths of a nanometer [167]. Considine et al. reported asperity heights of 20 nm on latex spheres of 6.4 and 7 µm in diameter [168]. Based on these measurements, roughness on one or both surfaces may have reduced the adhesion forces.

To support the results from the AFM force spectroscopy experiments, the experimental data was compared to theoretically calculated force-distance curves applying the DLVO theory. In the following text, the results obtained from the measurements and from theory are discussed with special attention to the different sample surfaces.

Adsorption and force spectroscopy results on single minerals

In the case of the colloid sorption experiments with background electrolyte and the **sheet silicates, albite, K-Feldspar, quartz and sapphire** (Figs. 24-28, 30) the decreasing fluorescence intensities with increasing pH can be explained by electrostatic repulsion. With increasing pH, the aluminol- and silanol-(edge) sites of the sheet silicates, feldspars, quartz (only silanol-sites), sapphire (only aluminol-sites) and the colloid carboxyl groups deprotonate resulting in increasing repulsive electrostatic forces. The pH_{pzc} values of muscovite and biotite surfaces were determined as 6.6 and 6.5, respectively [77,78]. Therefore, both the muscovite and biotite surfaces are negatively charged at alkaline pH values and repulsion towards negatively charged colloids dominates. This corresponds to the finding that on the permanent negatively charged sheet silicate basal planes no colloid adsorption was detected in the alkaline pH regime and in the absence of Eu(III). At pH 2, the surface charge of quartz, albite and K-feldspar is close to its pH_{pzc} (pH_{pzc} = 2 [52]; pH_{pzc} =

2.6 [169]) and the surface charge of the carboxylated latex particles is relatively low (~ -20 mV, see zeta-potential measurements in Fig. 19). The neutralization of surface charge (and thus overall reduction of repulsive interactions) resulted in maximum values for colloid adsorption. However, on albite and K-feldspar, weak fluorescence was detected even at pH 4, and in the case of albite also at pH 6, where both the mineral and colloid surface can be assumed to be negatively charged (Figs. 26 and 27). This observation cannot be explained by a simple electrostatic description.

According to Ryan and Elimelech [170] colloid sorption under repulsive conditions may be attributed to surface roughness and/or chemical heterogeneities. Surface non-idealities, such as cracks, dislocations, geochemical impurities may lead to a nonuniform charge distribution known as "surface charge heterogeneity" [171,172,173,174,175].

Note that colloid adsorption may also take place if only small repulsive forces exist. The smaller the energy barrier, the higher is the probability of colloid attachment.

Song et al. [176] developed a theoretical approach for the calculation of a colloid deposition rate onto heterogeneously charged surfaces. They found that even minor charge heterogeneities result in increased deposition rates compared with a homogeneous charge distribution. It is possible to gain insight into heterogeneously charged surfaces by carrying out AFM force-volume measurements.

Theoretical calculations by Adamczyk et al. showed increased colloid adsorption on solid surfaces due to surface roughness and chemical heterogeneities [177]. Hoek and Agarwal concluded from numerical simulations that rough surfaces are more favourable for colloid deposition due to long range (attractive) van der Waals interactions [178]. Therefore, it can be expected that colloid adsorption is strongly influenced by surface roughness, charge distributions and chemical heterogeneities, especially under electrostatically unfavourable conditions.

The colloid sorption experiments with **apatite** and background electrolyte showed that highest fluorescence intensity was found at low pH values and decreased with increasing pH (Fig. 29). But due to the comparatively high pH_{pzc} (about 7.6-8.1, see Table 1) colloid adsorption was observed even at pH 8. Again, the observed trends are in accordance with a simple electrostatic description of a positively charged mineral surface at pH values below the pH_{pzc} of negatively charged colloids.

Wu et al. [80] propose for fluorapatite $\equiv Ca-OH_2^+$ and $\equiv P-O^-$ as dominating surface groups which are characterized by the following equilibrium constants (10^{-1} M NaCl, 25°C):

V Discussion

$\equiv$P-O$^-$ + H$^+$ $\leftrightarrow$ P-OH   pK$_s$ = 6.6   (46)

$\equiv$Ca-OH$_2^+$ $\leftrightarrow$ Ca-OH + H$^+$ pK$_s$ = -9.7   (47)

Based on these equilibrium constants, the same authors expect for hydrous fluorapatite a pH$_{pzc}$ of 8.15: a value which is close to the pH$_{pzc}$ of hydroxyapatite mentioned above. Solubility data for hydroxyapatite show that increased dissolution of this mineral takes place with decreasing pH [179]. The dissolution reaction of hydroxyapatite, consuming protons, can be written as follows for acidic pH values [180]:

$$Ca_5(PO_4)_3OH + 7 H^+ \leftrightarrow 5 Ca^{2+} + 3H_2PO_4^- + H_2O \qquad (48)$$

The Ca release from the apatite (used in the sorption experiments) was determined by varying pH values from 2-8 at constant ionic strength (10^{-2} M NaCl) for 15 min. At pH 2 a weak dissolution of apatite during the experiment was observed resulting in a Ca concentration of about 5×10^{-5} M determined by ICP-MS (the Ca concentration of the corresponding stock solutions was not detectable). This low Ca concentration does not increase the ionic strength of the solution significantly and thus should not have a noticeable impact on colloid stability.

The force spectroscopy experiments on the **sheet silicates** could only be carried out on the basal planes of the sheet silicates as the mineral edges were not accessible in a defined way. In contrast to the sorption experiments, in the force spectroscopy experiments at low pH with background electrolyte attractive snap-in forces were detectable on the sheet silicates (Figs. 37 and 40). A reason for this contradiction may be the larger area of interaction of the larger colloid (~ 1 µm) compared to the smaller colloids (~ 25 nm) used in the sorption experiments. An effective contact area A between the 1 µm diameter colloid probe and a flat surface can be estimated to 943 nm² by using the Langbein approximation [183]: A=2πRx, where R is the colloid radius and x the separation distance. At contact, x can be estimated to 0.3 nm [183]. In contrast, the effective contact area of the fluorescent polystyrene colloids can be estimated to only 23 nm². Because of the larger interaction area it may be that attractive snap-in forces lead to colloid adsorption at low pH values (where the negative surface charge of the basal planes and colloids are already minimized). In agreement with the sorption experiments, no attractive snap-in forces were measured > pH 3 due to the increasing repulsion between the surfaces.

V Discussion   98

The force spectroscopy measurements with **K-feldspar** and **quartz** showed that, in the absence of additional cations, snap-in forces have maximum values at low pH corresponding with maximum fluorescence intensities in the sorption experiments (Figs. 42 and 45). The observation that snap-in forces were detected on K-feldspar even at pH 4 and that these forces were not detectable above this pH also agrees with the results of the sorption experiments. Since the pH_{pzc} of the quartz used in the force spectroscopy experiments was determined to 3.8 (see Fig. 21), this observation can be explained by electrostatic attraction.

The corresponding force measurements with **apatite** and background electrolyte contrasted somewhat with the sorption experiments: maximum snap-in forces were observed at pH 5 and with decreasing pH they also decreased, whereas in the sorption experiments the fluorescence increased with decreasing pH (Fig. 48). Above the pH_{pzc}, no attractive snap-in forces were measured, an observation which corresponds again with the sorption measurements. The decrease of attractive forces with decreasing pH is an unexpected result, since an increase of positive mineral surface charge is expected with decreasing pH. Furthermore, the colloid surface charge will become less negative with decreasing pH. It is known from the literature that with decreasing pH dissolution of the mineral surface takes place [181]. AFM contact mode images at different pH clearly prove the increasing surface roughness with decreasing pH (Fig. 50 and Fig. 51). It can be assumed that mineral dissolution results in a smaller effective contact area between the colloid and mineral surface. This in turn can significantly reduce the attractive van-der-Waals and adhesion forces [182,183,184,185]. For the smaller colloids the contact area plays a less significant role (as stated above, the effective contact area of the small colloids is significantly smaller than that of the large colloids) so that the surface-colloid interaction is influenced to a far lesser degree by a decreasing contact area. This may be an explanation why the force spectroscopy measurements show contradictory behaviour compared to the sorption experiments. Although different types of minerals were used in the sorption (hydroxyapatite) and force spectroscopy experiments (fluorapatite) the dissolution rates of both minerals are similar [186], so that a differing dissolution rate can be excluded as an explanation for the contradicting observations.

Influence of Eu(III)

Regarding the sorption experiments on the **sheet silicates** in the presence of Eu(III) a significant increase of fluorescence intensity both on mineral edges and planes was observed

V Discussion   99

(Figs. 24 and 25). Eu(III) also affected the colloid adsorption on the basal planes where no adsorption was observed in the absence of polyvalent cations. These observations are in agreement with a binding mechanism involving polyvalent cation bridges [187] and was also assumed in the case of colloid adsorption of negatively charged humic colloids on muscovite [188]. At the edges of both sheet silicates increased fluorescence intensities were found in the experiments with the higher Eu(III) concentration. Colloids can obviously adsorb under these conditions on the edge aluminol- and silanol groups although these groups are protonated at pH 2 [189,82]. The increase of colloid-mineral adsorption in the presence of polyvalent cations at low pH can not be described with this simple approach. Inner-sphere complexation to either carboxylic acid groups of colloids or surface hydroxyl groups of the minerals should not be relevant at pH 2 [160]. The colloid surface charge is not significantly influenced by the presence of Eu(III), even though Eu(III) is present in excess over the carboxylate groups (10^{-5} M Eu(III) vs. $6*10^{-6}$ eq/l COOH). This indicates no or only low Eu(III) adsorption on the colloids, which is in agreement with the literature [190]. It can be assumed that at low pH Eu(III) is enriched at mineral surfaces by physisorption or cation exchange against, e.g., H^+, K^+, Ca^{2+} or Al^{3+}. The enhancement of colloid adsorption may be explained by surface bound Eu(III) which induces high local charge density. The force spectroscopy experiments with the **sheet silicates** in the presence of Eu(III) showed comparatively increased snap-in forces at pH values ≥ 4 (Figs. 37 and 40). This corresponds to the sorption experiments carried out with the fluorescent colloids in which increased fluorescence was also seen in presence of 10^{-5} M Eu(III). In contrast to the sorption experiments, very small attractive snap-in forces could be measured on muscovite even at the lower (10^{-6} M) Eu(III) concentration. This observation speaks for a comparatively increased sensitivity of the force spectroscopy as compared to the sorption experiments with fluorescent dyed colloids. The latter method is restricted to the sensitivity of the fluorescence detector.

For **albite**, **K-feldspar** and **sapphire** a significant influence of Eu(III) on colloid adsorption was observed in the sorption experiments, which in the case of K-feldspar was dominated by the mineral edges (Figs. 26, 27, 30). The colloid adsorption experiments in the presence of Eu(III) showed that the influence of this cation strongly depends on pH. Due to the protonation of the involved surfaces there was no influence of Eu(III) observed at pH 2. At pH 4 and 6, the influence of Eu(III) is more significant, depending on the Eu(III) concentration. This corresponds to the force spectroscopy measurements with K-feldspar (Fig. 42), where a significant increase of the snap-in forces (dependent on Eu(III) concentration) at

V Discussion 100

pH ≥ 4 was observed. Blum and Lasaga [191] and Walther [192] state that, for feldspars, {> (Al, Si)-OH$_n^{n-1}$} groups at the surface are the dominant charged species which are potential binding sites for Eu(III). Similar findings were observed in the sorption experiments with **quartz** (Fig. 28). In the presence of the lower Eu(III) concentration no effect could be detected, although one can assume that Eu(III) adsorbs on the quartz surface at pH ≥ 4 [193]. This was supported by the fact, that in the presence of the higher Eu(III) concentration a weak colloid adsorption was detected at pH 6. Eu(III) binds to the silanol-groups on the quartz surface and thus increases the positive mineral surface charge. In analogy to the sorption experiments, snap-in forces were observed on quartz at pH 6 in the presence of Eu(III). But compared to the sorption experiments, the more sensitive force measurements detected snap-in forces also at pH 4 (Fig. 45).

The colloid sorption experiment with **apatite** in the presence of Eu(III) showed an increase of the measured fluorescence intensities only with high Eu(III) concentrations (Fig. 29). This observation might be explained by a complexation of Eu(III) with remaining negative complexing groups on the mineral surface, thus increasing the positive mineral surface charge. This idea is supported by the fact that the strongest colloid fluorescence was found at pH 6. The force spectroscopy experiments with **apatite** did not show significant influence of Eu(III) (pH 2-6) compared to the measurements in absence of these cations (Fig. 48). A clear explanation for the different observations made by sorption and force measurements cannot be given right now. For apatite the measured snap-in forces are highest as compared to all other minerals under investigation showing the strong affinity of the surface towards negatively charged colloids. One may thus assume that the impact of sorbing cations is of second order and does not significantly modify the apatite surface charge properties. Enhancement of forces from 0 to 0.2 nN are found for the other mineral surfaces due to addition of 10^{-5} M Eu(III). Such variation of forces might well lie within the error bars of forces measured for apatite where even in absence of Eu(III) snap-in forces lie at around 1 nN.

Influence of Ca(II), UO_2^{2+} and Grimsel groundwater

Ca(II) and UO_2^{2+} did not conceivably influence colloid adsorption to the investigated mineral surfaces in force spectroscopy. It can be assumed that these cations will adsorb onto the mineral surfaces at pH values ≥ 4 [194,195,196,197] but the resulting snap-in forces were not detectable in the experiments.

However, at pH 10, the adsorption of Ca(II) onto apatite (note that the pH_{pzc} is around 8, see Table 1) and the colloid may increase both surface potentials which leads to the slightly increased adhesion and snap-in forces.

The sorption and force spectroscopy experiments with **sheet silicates, albite** (only sorption experiment), **K-feldspar, quartz, sapphire** (only sorption experiment), **titanite** and Grimsel groundwater proved that significant colloid adsorption can not be expected since neither fluorescence nor snap-in forces were detected (Figs. 24-28, 30, 37, 40, 42, 45). Cations present in the groundwater may adsorb onto the involved surfaces but apparently still the repulsive forces are not sufficiently minimized to allow for (detectable) colloid adsorption under the experimental conditions. The experiments with Grimsel groundwater and **apatite** (Fig. 48) showed that both types of experiments complement each other well: snap-in forces and colloid adsorption were detected and thus colloid adsorption can be expected under natural conditions on this mineral.

DLVO calculations and adhesion forces

A more detailed view on the repulsive forces at pH 10 on **muscovite, K-feldspar and quartz** in presence and absence of Ca(II) did show that a significant decrease of repulsive forces takes place in presence of Ca(II) (Figs. 39, 44, 47). Since biotite has a similar mineralogical structure as muscovite, a similar behaviour can be assumed for both sheet silicates. The sorption of Ca(II) on the interacting surfaces possibly increases the individual surface potentials which leads to a reduction of repulsion by a factor of ~ 3. Three positions on the sheet silicate basal planes are available to cations: (1) as counter-ions in the diffuse layer, (2) as adsorbed hydrated ions at the Stern plane and (3) as dehydrated ions residing in the lattice cavities of the basal sheet [198]. An adsorption of these cations onto the surface of the sheet silicates at pH > 4 is likely but obviously the repulsive energy barrier can not be overcome by the colloids at least during the time period of the experiments. The observation, nevertheless, indicates that the probability of colloid sticking to the surface increases significantly and may lead to colloid attachment at longer contact times.

Force-distance curves with quartz and muscovite showed that experimental colloid-mineral surface interaction data can be well described by the DLVO theory calculations (Figs. 55-58). The fitted surface potentials of quartz correlated well with the measured zeta-potentials (Table 14) for quartz. The fitted surface potential values of muscovite at pH 6 were also close to the according zeta-potentials found in the literature (-35 - -40 mV according to ref. [199,200]; the DLVO fitted value was -30 mV). According to the DLVO fits, the adsorption of Eu(III) on the muscovite, quartz and colloid surfaces obviously significantly increases the attractive interaction potential. However, at separation distances < 5 nm a disagreement of experimental data and theory can be seen: in the experiments at pH 6 (background electrolyte) or Grimsel groundwater, theory predicts attractive van der Waals forces which were not observed in the experiments. The reason for this is the presence of repulsive hydration forces. Other authors also observed hydration forces on muscovite: Pashley and Israelachvili carried out a series of experiments to identify the factors that regulate hydration forces [201]. The experiments were carried out with KCl solutions with different concentrations. With 10^{-5} and 10^{-4} M KCl the force curve followed the theoretical DLVO force law at all separations. At 10^{-3} M KCl and higher concentrations more cations adsorb onto the surfaces and bring with them their water of hydration which gives rise to an additional short-range hydration force. This is due to the energy needed to dehydrate the bound cations which presumably retain some of their water of hydration on binding. In acidic solutions, where only protons bind to the surfaces, no hydration forces were observed (the authors suggest that is because protons penetrate into the mica lattice) and the measured force data was close to the theoretical prediction.

Similar conclusions can also be assumed for the observations made with biotite (Fig. 59) since this mineral has a similar mineralogical composition as muscovite.

Other investigators also observed hydration forces on quartz which are believed to arise from strongly bonding H-bonding surface groups, such as hydrated ions or hydroxyl groups, which modify the H-bonding network of liquid water adjacent to them [91].

The DLVO calculations with apatite (Fig. 60) showed that the apatite surface obviously becomes neutralized under Grimsel groundwater conditions which leads to the observed attractive forces.

Maximum repulsive barrier calculations with selected theoretical calculated force curves showed that a certain colloid attachment probability can be expected on the minerals even when generally repulsive conditions are apparent. Although the calculations were only performed for the theoretical calculated force curves they still may give an insight into colloid attachment probability of the experimental measurements.

The measured adhesion forces on the **sheet silicates, K-feldspar and quartz** decreased with increasing pH in the presence of background electrolyte (Figs. 38, 41, 43, 45). This observation is in line with snap-in force and sorption measurements. It is however remarkable that the presence of Eu(III) and Ca(II) lead to significantly enhanced adhesion forces even under those conditions where no snap-in forces are detected. As already discussed above, the exact meaning of the measured adhesion forces for colloid attachment to surfaces is not clear. Increasing adhesion forces are observed going along with decreasing repulsive forces. Differences between snap-in forces and adhesion forces lie in the order of magnitude of ~1 nN. This is quite similar to what is assumed as a typical value for hydration forces in extended DLVO model approaches: 1.6 nN/nm^2 [202] assuming a contact area of about 943 nm^2. In this case the snap-in forces can be taken as a measure for colloid attraction to the surface. If the colloid manages to cross the hydration barrier, the adhesion forces give a measure for the "binding" strength of the colloid to the surface. This colloid "binding" strength can be taken to assess colloid desorption from the surface via shear forces exerted by groundwater flow (see discussion below).

Another factor that could influence the colloid-mineral interaction may be the presence of steric forces. Since polystyrene spheres were used one has to assume that the surface of these polymers will not be smooth but rather will have a "hairy" nature (and thus thermally diffuse interface) as a result of protruding polymer chains [203]. The polymer chains extend out into the solution and are there thermally mobile. On approach of another surface the entropy of confining of these dangling chains again results in a repulsive force known as steric repulsion. Seebergh and Berg [204] carried out experiments with polystyrene colloids to verify the presence of the "hairy" layer. In their study, the presence of this layer was investigated by comparing size, mobility, critical coagulation concentration and surface charge density measurements and also after heat treatment. The results showed that the hydrodynamic diameter of the particles is strongly dependent on electrolyte concentration: the hydrodynamic diameter decreases with increasing electrolyte concentration (from 10^{-6} M to 10^{-2} M KNO$_3$/NaCl) due to the gradual collapse of the polymer chains or "hairs". It was shown that heat treatment eliminates the hairs and smoothes the surface. The measured hydrodynamic diameters of the heat treated particles was equal in size to the untreated colloids investigated at 10^{-3} and 10^{-2} M NaCl (at lower ionic strengths, the polymer chains are extended to a higher degree hence increasing the hydrodynamic diameter; the heat treated particles had the same hydrodynamic diameter at all ionic strengths). According to the authors, this result was consistent with their hypothesis that particles have a hairy layer which collapses upon addition of electrolyte and is completely removed on heat treatment.

Since the experiments carried out in this work were carried out with ionic strengths of 10^{-3} M (Grimsel groundwater) and 10^{-2} M NaCl it can be assumed that the polymer "hairs" are collapsed and thus the steric force should be negligible.

Note that it is not certain that a hairy layer on the polymer particles used in this work even exists. The surface structure of the particles strongly depends on the manufacturing process. The manufacturer of the colloids confirmed that the surface of the colloids used in the AFM force spectroscopy experiments is smooth (as was seen with SEM) and does not possess any (with SEM detectable) protruding polymer chains at all. In this case no influence of steric forces would be expected.

Force-volume measurements

In the force-volume measurements no significant deviations of the snap-in or adhesion forces from those obtained in the single point measurements were observed. It can be assumed that the size of the colloidal particle plays an important role because the net force acting between colloid and surface depends on the interaction area. As stated above, the effective contact area between colloid probe and flat surface can be estimated to around 943 nm². Possible variations of surface charge will be averaged over this relatively large contact area. Thus, a possible effect of surface charge heterogeneities under all measurement conditions will depend on the relative size of the particle with respect to the relative size of the heterogeneously charged regions.

Grimsel Granodiorite

In the Grimsel granodiorite system, colloid sorption mainly takes place on the minerals apatite, illite, titanite, biotite (pH 2-6) and K-feldspars (pH 2-4). No adsorption could be detected at alkaline pH in measurements with background electrolyte only and with Grimsel groundwater. In the presence of Eu(III) increased colloid sorption on granodiorite was detected. These observations agree with those made with the single minerals. The desorption experiment with Grimsel Granodiorite showed that colloid adsorption is reversible. This was expected because in the corresponding force spectroscopy experiments with pH = 10 and I = 10^{-2} M NaCl only repulsive interaction was detected on the Grimsel granodiorite main component minerals. Furthermore, in the colloid adsorption experiments under the same conditions no colloid adsorption was found.

Colloid desorption from mineral surfaces is controversially discussed in the literature. Ryan and Gschwend carried out colloid desorption experiments and measured the rates of clay colloid release from iron oxide coated sand. The experiments indicated that the release of the colloids was controlled by the changes in electrostatic interactions between colloid surfaces resulting from perturbations of solution chemistry. They found that with increasing pH (and thus repulsion between the surfaces) and decreasing ionic strength the rates of colloid detachment increased. At lower ionic strength, the repulsive double layer extends further into solution [205]. However, Kallay et al. [206] observed an increase in release rate when the ionic strength was increased in a system of hematite colloids and glass grains. This contradictory result was attributed to the use of a shorter column that supposedly reduced the probability of reattachment of released colloids that would normally be promoted at high ionic strengths [207]. Fogler et al. found that colloid release occurred suddenly when ionic strength or [H^+] reached a threshold beyond which colloids were mobilised [208,209,210]. In these and other experiments colloid release occurred under conditions for which DLVO calculations predicted that detachment energy barriers were still present [211,212]. Ryan and Gschwend hypothesised that the threshold values responsible for rapid colloid mobilisation actually correspond to the conditions at which the detachment energy barriers disappear and the rates of colloid release are no longer limited by detachment from the surface [205].

To conclude, the results of the desorption experiment carried out in this work correspond to the concept that under generally repulsive conditions a colloid desorption can be expected. This is comparable to the observations made in the detachment tests of Ryan and Gschwend (mentioned above) who observed increasing colloid release with increasing repulsive conditions, i.e., increasing pH.

To assess whether the adsorbed colloids might be sheared off the mineral surfaces by groundwater flow an approximation of the shear forces generated by water flow was developed by Nasr-el Din [213]: $F_D = 1.7009(6\pi\eta R v_x)$ where R is the colloid radius, η the dynamic viscosity and v_x the fluid velocity. Under natural conditions, groundwater flow velocities of about 1-10 m/a can be expected, whereas in the Grimsel system the groundwater flow velocity is about 1 m/a [218]. Shear forces in the range of $8*10^{-7}$ nN – $8*10^{-6}$ nN were estimated for flow velocities of 1 m/a and 10 m/a, respectively. Since all measured adhesion forces were significantly higher than the estimated shear forces (in the nN range), it can not be

V Discussion

expected that the colloids, once adsorbed, will be remobilised from the surface due to ground water flow. Even with the artificially generated higher flow velocities in the CRR project in the GTS [8] (see also section I) no colloid mobilisation can be expected. Estimated shear forces were here in the range of 10^{-2} nN, which is still slightly lower than the lowest experimental adhesion forces. Other authors who investigated the release of colloids under laminar flow also found that hydrodynamic shear forces are negligible compared to electrostatic adhesive forces [214,215].

VI Summary and Conclusions

To sum up, the results of this study show that the interaction of colloids with mineral surfaces is mainly controlled by electrostatic interactions. The experiments with background electrolyte show that the strongest attractive forces/fluorescence intensity are/is observed close to or below the individual pH_{pzc} of the minerals. The highest adhesion forces with background electrolyte were observed in the experiments with muscovite and biotite followed by apatite, K-feldspar and finally quartz. Regarding the snap-in forces, the highest values were observed on apatite, muscovite, biotite, K-feldspar and quartz. Thus, it can be concluded that apatite and the sheet silicates are most prone to colloid adsorption, whereas on quartz the least adsorption of colloids can be expected.

According to the results, rapid colloid adsorption in the alkaline regime (e.g., Grimsel groundwater conditions) can only be expected on apatite in the presence of Ca(II) or with Grimsel groundwater. Since apatite occurs only in trace concentrations in the Grimsel Granodiorite, only negligible colloid retention (due to electrostatic interaction) would be expected in the Grimsel system. This is in agreement with the laboratory and field experiments carried out by Geckeis et al. [7], Möri et al. [8] and Kosakowsky [216] where clay colloids were only insignificantly retarded. However, these experiments were carried out with relatively high water flow rates and are thus not fully representative of the water flow rates expected in a deep geological repository. Other studies showed significantly lower colloid recoveries in laboratory and field studies [9],[217]. Degueldre et al. [218] carried out sorption experiments at pH 8 (I = 10^{-2} M NaClO$_4$) with clay colloids and several single minerals (muscovite, biotite, quartz, feldspar) comprising the Grimsel granodiorite and also observed considerable colloid adsorption.

Alonso et al. [219] studied the adsorption of Au colloids on granite surfaces under alkaline (pH 9.5) conditions with µPIXE. Despite the expected repulsion an Au colloid retention was observed on the granite surface. It was found that colloids were sometimes locally retained on Ca and Fe bearing minerals. It was suggested that the presence of bivalent ions like Ca at the surface might promote local aggregation of colloids, because dissolved Ca may enhance particle aggregation and their retention. This observation corresponds to the observation made in this work that sorption takes place on the Ca-phase apatite. However, no colloid adsorption at alkaline pH could be detected on Fe-phases in this work. The reason for this is (besides that sorption may not take place) that fluorescence of dyed colloids may be quenched by Fe and thus can not be detected. No sufficiently smooth samples could be acquired for AFM force spectroscopy experiments on Fe-phases. Furthermore, the granite used in the experiments of Alonso et al. [219] has a higher content of Fe-phases. The Grimsel granodiorite fracture filling material used in this work has only traces or very small concentrations of Fe-phases (ilmenite, orthite, chlorite, epidote) and hence they possibly do not play a significant role regarding a possible colloid retention.

Several bentonite colloid migration experiments carried out by Seher [220], on the other hand, showed unexpectedly high colloid recoveries (70-100 %) in laboratory and field experiments even with very slow (natural) flow rates. Long-term batch experiments (1000 hours) carried out with bentonite colloids and Grimsel granodiorite in Grimsel groundwater also did not show any significant colloid adsorption [221].

If electrostatic interaction can be ruled out as the origin of colloid retention under repulsive conditions, other mechanisms for colloid retention must exist. Missana et al. [222] studied the dependence of colloid recovery on colloid concentration and water flow rate in a Grimsel granodiorite core. At very low flow rates and low colloid concentrations the colloid recovery was near 100 %, whereas decreased recoveries of only 20 % with high colloid concentration were found. The authors concluded that an increase in the viscosity of the clay suspensions with increasing colloid concentration could be the cause of stronger particle-particle interactions leading to ripening effects: a multi-layer adsorption of the colloids onto the surface of the stationary phase may take place. As a consequence, the deposition rate of colloids increases with increased number of adsorbed particles.

Increased colloid retention under repulsive conditions may also occur due to the influence of mineral surfaces roughness. Ryan and Elimelech [223] reported higher than predicted colloid-rock attachment under repulsive conditions and hypothesised the role of the surface physical heterogeneities, i.e., surface roughness.

To summarise future research priorities, further studies are necessary to understand colloid filtration under repulsive conditions, especially the role of physical heterogeneity, e.g., by carrying out colloid transport or adsorption experiments on (e.g., chemically inert) surfaces with defined surface roughness or imperfections.

Furthermore, the presence and dimensions of surface charge heterogeneities should be investigated by electric force microscopy. With this method it is possible to map the sample surface potential with near atomic resolution. Theoretical calculations could estimate the degree of influence on colloid-collector interaction of possibly present heterogeneously charged "patches".

V Discussion

By using vertical laser scanning interferometry the adsorption of colloidal particles with diameters of at least ~ 500 nm can detected with high resolution. This method could be applied for sorption experiments with natural bentonite colloids on relevant mineral samples. To further develop the DLVO calculations, contact angle measurements and Hamaker constant calculations should be carried out with the polystyrene colloids in order to be independent of literature values. A more exact and reliable (but very elaborate) way to determine the Hamaker constants (on the basis of the Lifshitz theory) would be to measure the dielectric functions for all frequencies of the solids.

VII Literature

[1] Kim, J.I., Gompper, K., Geckeis, H., Forschung zur Langzeitsicherheit der Endlagerung radioaktiver Abfälle, Radioaktivität und Kernenergie, S. 118, Mai 2001.

[2] Alonso, U., Missana,T., Geckeis, H., Garcia-Gutierrez, M., Turrero, M.J., Möri, R., Schäfer, Th., Patelli, A., Rigato, V., 2006. Role of inorganic colloids generated in a high-level deep geological repository in the migration of radionuclides: Open questions. Journal of Iberian Geology 32 (1), 79.

[3] Yoshida, H., Takeuchi, M., Metcalfe, R., 2005. Long-term stability of flow-path structure in crystalline rocks distributed in an orogenic belt, Japan. Engineering Geology 78, 275.

[4] Ryan, J.N., Gschwend, P.M., 1994. Effects of ionic strength and flow rate on colloid release: Relating kinetics to intersurface potential energy. Journal of Colloid and Interface Science 164, 21.

[5] Gallé, C., 2000. Gas breakthrough pressure in compacted Fo-Ca clay and interfacial gas overpressure in waste disposal context. Applied Clay Science 17, 85-97.

[6] Kersting, A.B., Efurd, D.W., Finnegan, D.L., Rokop, D.J., Smith, D.K., Thompson, J.L., 1999. Migration of Plutonium in ground water at the Nevada Test Site. Nature 397, 56.

[7] Geckeis, H., Schäfer, T., Hauser, W., Rabung, Th., Missana, T., Degueldre, C., Möri, A.,Eikenberg, J., Fierz, Th., Alexander, W.R., 2004. Results of the colloid and radionuclide retention experiment (CRR) at the Grimsel Test Site (GTS), Switzerland – impact of reaction kinetics and speciation on radionuclide migration. Radiochimica Acta 92, 765.

[8] Möri, A., Alexander, W.R., Geckeis, H., Hauser, W., Schäfer, T., Eigenberg, J, Fierz, Th. Degueldre, C., Missana, T., 2003. The colloid and radionuclide retardation experiment at the Grimsel Test Site: influence of bentonite colloids on radionuclide migration in fractured rock. Colloids and Surfaces A: Physicochemical and Engineering Aspects 217, 33.

[9] Schäfer, T., Geckeis, H., Bouby, M, Fanghänel, T., 2004. U, Th, Eu and colloid mobility in a granite fracture under near-natural flow conditions. Radiochimica Acta 92, 731.

[10] Missana, T., Alonso, U., Garcia-Gutiérrez, M., Mingarro, M., 2008. Role of bentonite colloids on europium and plutonium migration in a granite fracture. Applied geochemistry 23, 1484.

[11] Patelli, A., Alonso, U., Rigato, V., Missana, T., Restello, S., 2006. Validation of the RBS analysis for the colloid migration through a rough granite surface. Nuclear Instruments and Methods in Physics Research B 249, 575.

[12] Alonso, U., Missana, T., Patelli, A., Rigato, V., Ravangan, J., 2007. Colloid diffusion in crystalline rock: An experimental methodology to measure diffusion coefficients and evaluate colloid size dependence. Earth and Planetary Science Letters 259, 372.

[13] Stumpf, S., Stumpf, Th., Lützenkirchen, J., Walter, C., Fanghänel, Th., 2008. Immobilization of trivalent actinides by sorption onto quartz and incorporation into siliceous bulk: Investigations by TRLFS. Journal of Colloid and Interface Science 318, 5.

[14] Vakarelski, I.V., Ishimura, K., Higashitani, K., 2000. Adhesion between Silica Particle and Mica Surfaces in Water and Electrolyte Solutions. Journal of Colloid and Interface Science 227, 111.

[15] Bowen, W.R., Doneva, T.A., 2000. Atomic Force Microscopy Studies of Membranes: Effect of Surface Roughness on Double-Layer Interactions and Particle Adhesion. Journal of Colloid and Interface Science 229, 544.

[16] Johnson, P.R., Sun, N., Elimelech, M., 1996. Colloid Transport in Geochemically Heterogeneous Porous Media: Modeling and Measurements. Environmental Science and Technology 30, 3284.

[17] Alonso, U., Missana, T., Patelli, A., Rigato, V., Ravagnan, J., 2004. μPIXE study on colloid heterogeneous retention due to colloid/rock electrostatic interactions. Laboratori Nazionali di Legnaro Annual Report 2004, 68.

[18] Velegol, D., Thwar, P., 2001. Analytical Model for the Effect of Surface Charge Nonuniformity on Colloidal Interaction. Langmuir 17, 7687.

[19] Taboada-Serrano, P., Vithayaveroj, V., Yiacoumi, S., Tsouris, C., 2005. Surface Charge Heterogeneities Measured by Atomic Force Microscopy. Environmental Science and Technology 39, 6352.

[20] Dörfler, H.-D., 2002. Grenzflächen und kolloid-disperse Systeme: Physik und Chemie. Springer-Verlag, Berlin, Heidelberg, New York.

[21] Hirner, A. V., Rehage, H., Sulkowski, M., 2000. Umweltgeochemie – Herkunft, Mobilität und Analyse von Schadstoffen in der Pedosphäre. Steinkopff Verlag, Darmstadt.

[22] Buffle, J., Perret, D., Newman, M.: The use of filtration and ultrafiltration for size fractionation of aquatic particles, colliods and macromolecules. In: Buffle, J., van Leeuwen, H.P.: Environmental particles Volume 1. Environmental Analytical and Physical Chemistry Series. Lewis Publishers, Boca Raton, 171-230, 1992.

[23] Sposito, G., 1998. Bodenchemie. Ferdinand Enke Verlag, Stuttgart.

[24] Manual of symbols and terminology for physicochemical quantities and units. Appendix II: Definitions, terminology and symbols in colloid and surface chemistry. Part I. IUPAC,1972, http://www.iupac.org/reports/2001/colloid_2001/manual_of_s_and_t.pdf).

[25] McCarthy, J.F., Zachara, J.M., 1989. Subsurface Transport of Contaminants, Environmental Science and Technology 23, 496.

[26] Vilks, P., Cramer, J.J., Shewchuk, T.A., Larocque, J.P.A., 1988. Colloid and Particulate Matter Studies in the Cigar Lake Natural-Analog Program, Radiochimica Acta 44/45, 305.

[27] Yarin, S., Cross, H., 1979. Geochemistry of Colloids, Springer-Verlag.

[28] Degueldre, C.A., 1990. Grimsel Colloids Exercise, CEC Report EUR 12660 EN, Brussels.

[29] Kim, J.I., Buckau, G., Zhuang, W., 1986. Experimentelle Huminstoffuntersuchungen an Gorleben-Grundwässern, Institut für Radiochemie der TU München, RCM 03786.

[30] Kim, J.I., Buckau, G., Klenze, R., 1987. Natural Colloids and Generation of Actinide Pseudocolloids in Groundwater. In: Natural Analogues in Radioactive Waste Disposal (B. Come, N. Chapman, eds.). Graham and Trotman, London.

[31] Schachtschabel P., Scheffer, F., 1998. Lehrbuch der Bodenkunde, Ferdinand Enke Verlag, Stuttgart.

[32] Lövgen, L., Sjöberg, S., Schindler, P.W., 1990. Acid/base reactions and Al(III) complexation at the surface of goethite. Geochimica et Cosmochimica Acta 54: 1301.

[33] Gao, Y., Mucci, A., 2003. Individual and competitive adsorption of phosphate and arsenate on goethite in articifical seawater. Chemical Geology 199, 91.

[34] Goldberg, S., Lesch, S. M., Suarez, D. L., Basta, N. T., 2005. Predicting arsenate adsorption by soils by using soil chemical parameters in the constant capacitance model. Soil Science Society of America Journal 69, 1389.

[35] Christl, I., Kretschmar, R., 1999. Competitive sorption of copper and lead at the oxide-water interface: Implications for surface site density. Geochimica et Cosmochimica Acta 63, 2929.

[36] Piasecki, W., 2002. 1pK and 2pK protonation models in the theoretical description of simple ion adsorption at the oxide/electrolyte interface: A comparative study of the predicted and observed enthalpic effects accompanying adsorption of simple ions. Langmuir 18: 4809.

[37] Hofmann, A., van Beinum, W., Meeussen, J.C.L., Kretzschmar, R., 2005. Sorption kinetics of strontium in porous hydrous ferric oxide aggregates II. Comparison of experimental results and model predictions. Journal of Colloid and Interface Science 283, 29.

[38] Zarzycki, P., 2007. Monte Carlo modelling of ion adsorption at the energetically heterogeneous metal oxide/electrolyte interface: Micro- and macroscopic correlations between adsorption energies. Journal of Colloid and Interface Science 306, 328.

[39] Brinkmann, A.G., 1993. A double-layer model for ion adsorption onto metal oxides, applied to experimental data and to natural sediments of Lake Veluwe, The Nederland. Hydrobiologica 253, 31.

[40] Wang, F., Chen, J., Forsling, W., 1997. Model sorption of trace metals on natural sediments by surface complexation model. Environmental Science and Technology 31, 448.

[41] Kitamura, A., Fujiwara, K., Yamamoto, T., Nishikawa, S., Moriyama, H., 1999. Analysis of adsorption behaviour of cations onto quartz by electrical double layer model. Journal of Nuclear Science and Technology 36, 1167.

[42] Tonkin, J. W., Balistrier, L. S., Murray, J.W., 2004. Modeling sorption of divalent metal cations on hydrous manganese oxide using the diffuse double layer model. Applied Geochemistry 19, 29.

[43] Rudzinski, W., Charmas, R., Piasecki, W. 1999. Searching for thermodynamic relations in ion adsorption at oxide/electrolyte interfaces studied by using the 2-pK protonation model. Langmuir 15, 8553.

[44] Leroy, P., Revil, A., 2004. A triple-layer model of the surface electrochemical properties of clay minerals. Journal of Colloid and Interface Science 270, 371.

[45] Sverjensky, D.A., 2005. Prediction of surface charge on oxides in salt solutions: Revisions for 1:1 (M+L-) electrolytes. Geochimica et Cosmochimica Acta 69, 225.

[46] Piasecki, W., 2006. Determination of parameters for the 1-pK triple-layer model of ion adsorption onto oxides from known parameter values for the 2-pK TLM. Journal of Colloid and Interface Science 302, 389.

[47] Hiemstra, T., van Riemsdijk, W. H., 1999. Surface structural ion adsorption modelling of competitive binding of oxyanions by metal (hydr)oxides. Journal of Colloid and Interface Science 210: 182.

[48] Rietra, R. P. J.J., Hiemstra, T., van Riemsdijk, W. H., 2001. Comparison of Selenate and Sulfate Adsorption on Goethite. Journal of Colloid and Interface Science 240, 384.

[49] Ponthieu, M., Juilllot, F., Hiemstra, T., van Riemsdijk, W.H., Benedetti, M.F., 2006. Metal ion binding to iron oxides. Geochim. Cosmochim. Acta 70: 2679-2698.

[50] Van Riemsdijk, W.H., Weng, L., Hiemstra, T., 2007. Ion – colloid – colloid interactions. In: Frimmel, F.H., von der Kammer, F., Flemming, H.-C.: Colloidal transport in porous media. Springer-Verlag, Berlin, Heidelberg, New York.

[51] Scheffer, F., Schachtschabel, P., 2002. Lehrbuch der Bodenkunde. Spektrum Akademischer Verlag GmbH, Heidelberg, Berlin.

[52] Stumm, W., 1992. Chemistry of the solid-water interface. Processes at the mineral-water and particle-water interface in natural systems. John Wiley & Sons, New York.

[53] McCarthy, J. F., Zachara, J. M., 1989. Subsurface transport of contaminants. Environmental Science and Technology 23, 496.

[54] Ryan, J. N., Elimelech, M., 1996. Colloid mobilization and transport in groundwater. Colloids and Surfaces A 107, 1.

[55] Kretzschmar, R., Borkovec, M., Grolimund, D., Elimelech, M., 1999. Mobile subsurface colloids and their role in contaminant transport. Advances in Agronomy 66, 121.

[56] Grolimund, D., Borkovec, M., Barmettler, K., Sticher, H., 1996. Colloid-facilitated transport of strongly sorbing contaminants in natural porous media: a laboratory column study. Environmental Science and Technology 30, 3118.

[57] Roy, S.B., Dzombak, D.A., 1997. Chemical factors influencing colloid facilitated transport of contaminants in porous media. Environmental Science and Technology 31: 656.

[58] Saiers, J.E., Hornberger, G.M., 1999. The influence of ionic strength on the facilitated transport of cesium by kaolinite colloids. Water Resources Research 35, 6, 1713.

[59] Denaix, L., Semlali, R. M., Douay, F., 2001. Dissolved and colloidal transport of Cd, Pb, and Zn in a silt load soil affected by atmospheric industrial deposition. Environmental Pollution 113, 29.

[60] Zhuang, J., Flury, M., Jin, Y., 2003. Colloid-facilitated Cs transport through water saturated Hanford sediment and Ottawa sand. Environmental Science and Technology 37, 4905.

[61] Chen, G., Flury, M., Harsh, J.B., Lichtner, P.C., 2005. Colloid-facilitated transport of cesium in variably saturated Hanford sediments. Environmental Science and Technology 39, 3435.

[62] Elimelech, M., O'Melia, C. R., 1990. Kinetics of deposition of colloidal particles in porous media. Environmental Science and Technology 24, 1528.

[63] Kretzschmar, R., Sticher, H., 1997. Transport of humic coated iron oxide colloids in a sandy soil: Influence of Ca^{2+} and trace metals. Environmental Science and Technology 31: 3497.

[64] Kretzschmar, R., Barmettler, K., Grolimund, D., Yan, Y., Borkovec, M., Sticher, H., 1997. Experimental determination of colloid deposition rates and collision efficiencies in natural porous media. Water Resources Research 33, 5.

[65] Liu, D., Johnson, P.R., Elimelech, M., 1995. Colloid deposition dynamics in flow through porous media: Role of electrolyte concentration. Environmental Science and Technology 29, 2963.

[66] Ryde, N., Kallay, N., Matijevic, E., 1991. Particle adhesion in model systems. Part 14. – Experimental evaluation of multilayer deposition. Journal of the Chemical Society, Faraday Transaction Articles 87,9, 1377.

[67] Kulkarni, P., Sureshkumar, P., Biswas, P., 2005. Hierarchical approach to model multilayer colloidal deposition in porous media. Environmental Science and Technology 39, 6361.

[68] Litton, G.M., Olson, T.M., 1993. Colloid deposition rates on silica bed media and artifacts related to collector surface preparation methods. Environmental Science and Technology 27. 185.

[69] Christl, I., Kretzschmar, R., 1999. Competitive sorption of copper and lead at the oxide-water interface: Implications for surface-site density. Geochimica et. Cosmochimica Acta 63, 19/20, 2929.

[70] Christl, I., Kretzschmar, R., 2001. Interaction of copper and fulvic acid at the hematite-water interface. Geochimica et. Cosmochimica Acta 65, 3435.

[71] Venema, P., Hiemstra, T., van Riemsdijk, W. H., 1996. Comparison of different site binding models for cation sorption: Description of pH dependency, salt dependency and cation proton exchange. Journal of Colloid and Interface Science 181, 45.

[72] Abollino, O., Aceto, M., Malandrino, M., Sarzanini, C., Mentasti, E., 2003. Adsorption of heavy metals on Na-montmorillonite. Effect of pH and organic substances. Water Resources Research 37: 1619.

[73] Sparks, D.L., 2003. Environmental Soil Chemistry. Academic Press, London.

[74] Alexander, W.R., Ota K. & Frieg, B., 2001. The NAGRA-JNC in situ study of safety relevant radionuclide retardation in fractured crystalline rock II: the RRP project methodology development, field and laboratory tests. NAGRA Technical Report Series NTB 00-06, NAGRA, Wettingen, Switzerland.

[75] Stumm, W., 1992. Chemistry of the solid-water interface, John Wiley and Sons Inc., New York.

[76] Stumpf, S., Stumpf, Th., Walther, C., Bosbach, D., Fanghänel, Th., 2006. Sorption of Cm(III) onto different Feldspar surfaces: a TRLFS study. Radiochimica Acta 94, 243-248.

[77] Sverjensky, D.A., 1994. Zero-point-of-charge prediction of crystal chemistry and solvation theory. Geochimica et Cosmochimica Acta 58, 3123-3129.

[78] Alonso, U., 2003. Influencia de los coloides en el transporte de contaminantes en la interfaz campo cerano/campo lejano de un almacenamiento de residuos radioactivos de alta actividad. Ph.D. Thesis, Universidad Complutense de Madrid, Spain. In Spanish.

[79] Kosmulski, M., 2001. Chemical properties of material surfaces. Marcel Dekker Inc., New York.

[80] Wu, L., Forsling, W., Schindler, P.W., 1991. Surface complexation of calcium minerals in aqueous solution. 1. Surface protonation at fluorapatite-water interfaces. Journal of Colloid and Interface Science 147, 178-185.

[81] Elimelech, M., Gregory, J., Jia, X., Williams, R.A., 1995: Particle Deposition & Aggregation. Measurement, Modelling and Simulation. Butterworth-Heinemann, Woburn, MA, USA.

[82] Evangelou, V.P., 1998. Environmental soil and water chemistry. John Wiley and Sons Inc., New York.

[83] Dzombak, D.A., Morel, F.M.M., 1990. Surface Complexation Modelling. John Wiley and Sons Inc., New York.

[84] Frens, G., Overbeek, J. Th. G., 1972. Repetition and the theory of electrostatic colloids. Journal of Colloid and Interface Science 38, 376.

[85] Rajagopalan, R., Kim, J.S., Adsorption of Brownian Particles in the Presence of Potential barriers, 1981. Effects of Different modes of Double Layer interaction. Journal of Colloid and Interface Science 83, 428.

[86] Gregory, J., 1975. Interaction of unequal double layers at constant charge. Journal of Colloid and Interface Science 51, 44.

[87] Hamaker, H.C, 1937. The London-van der Waals Attraction between Spherical Particles, Physica IV/10, 1058.

[88] Lifshitz, E. M., 1956. The Theory of Molecular Attractive Forces between Solids, Journal of Experimental and Theoretical Physics 28, 73.

[89] Johnson, K.L., Kendall, K., Roberts, A.D., 1971: Surface Energy and the Contact of elastic solids. Proceedings of the Royal Society London 324, 301.

[90] Derjaguin, B. V., Muller, V.M., Toporov, Y.P., 1975. Effect of Contact Deformations on the Adhesion of Particles, Journal of Colloid and Interface Science 53/2, 314.

[91] Israelachvili, J.N., 1992. Intermolecular and Surface Forces, Academic Press, London.

[92] Butt, H.-J., Graf, K., Kappl, M., 2006. Physics and Chemistry of Interfaces, 2. Edition, Wiley VCH, Weinheim.

[93] Visser, J., 1972. On Hamaker Constants: A Comparison between Hamaker constants and Lifshitz-van der Waals constants. Advances in Colloid and Interface Science 3, 331.

[94] Bergström, L. 1997. Hamaker Constants of Inorganic Materials. Advances in Colloid and Interface Science 70, 125.

[95] Hiemenz, P.C., Rajagopalan, R., 1997. Principles of Colloid and Surface Chemistry, 3rd Edition, Marcel Dekker Verlag, New York.

[96] Eber, M., 2004. Wirksamkeit von nanoskaligen Fließregulierungsmitteln, Dissertation, Julius-Maximilians-Universität Würzburg.

[97] Gregory, J., 1975. Interaction of unequal double layers at constant charge. Journal of Colloid and Interface Science 51, 44.

[98] Butt, H.-J., Cappella, B., Kappl, M., 2005. Force measurements with the atomic force microscope: Technique, interpretation and applications. Surface Science Reports 59, 1.

[99] Schäfer, T., Vorlesungsskript "Transport von Kolloiden im Untergrund", Blockkurs 24164 (Vorlesung & Übung), Freie Univerisät Berlin, Fachereich Geowissenschaften, 15.-19.1.2002.

[100] Sposito, G., 1998. Bodenchemie. Ferdinand Enke Verlag, Stuttgart.

[101] Frimmel, F.H., 1986: Scriptum „Wasserchemie für Ingenieure" zur Vorlesung Chemische Technologie des Wassers, Universität Karlsruhe, ZfGW-Verlag GmbH, Frankfurt/Main, Deutschland.

[102] Müller, R. H., 1996. Zetapotential und Partikelladung in der Laborpraxis: Einführung in die Theorie, praktische Meßdurchführung, Dateninterpretation. Wissenschaftliche Verlagsgesellschaft mbH Stuttgart.

[103] Ruckenstein, E., Prieve, C., 1976. On Reversible Adsorptionof Hydrosols and Repeptization. American Institute of Chemical Engineers Journal 22, 276.

[104] Frant, M., et al., 1999. Oberflächenenergetische und zellbiologische Charakterisierung keramischer Biomaterialien des ternären Systems Al_2O_3-SiO_2-TiO_2, Materialwissenschaft und Werkstofftechnik 30/1, 24.

[105] Götzinger, M., 2005: Zur Charakterisierung von Wechselwirkungen partikulärer Feststoffoberflächen, Dissertation, Universität Erlangen – Nürnberg.

[106] Churaev, N. V., 1995. Contact Angles and Surface Forces, Advances in Colloid and Interface Science 58, 87.

[107] Erbil, Y., 2006. Surface Chemistry of Solid and Liquid Interfaces. Blackwell Publishing Ltd., Oxford, UK.

[108] van Oss, C. J., 1994. Interfacial Forces in Aqueous Media. Marcel Dekker, New York.

[109] Kwok, D. Y., Neumann, A. W. (xxxx): Contact Angle measurements and contact angle interpretation. Advances in Colloid and Interface Science, accepted.

[110] Fox, H.W., Zisman, W.A., 1952. The spreading of liquids on low energy surfaces, III. Hydrocarbon surfaces, Journal of Colloid Science 7, 428.

[111] McNeil-Watson, F., Tscharnuter, W., Miller, J., 1998. A new instrument for the measurement of very small electrophoretic mobilities using phase analysis light scattering (PALS). Colloids Surf., A 140, 53.

[112] Hiemenz, P.C., 1986. Principles of Colloid and Surface Chemistry, Marcel Dekker, New York.

[113] Müller, R.H., Schuhmann, R., 1996, Teilchengrößenmessung in der Laborpraxis, Wissenschaftliche Verlagsgesellschaft mbH, Deutschland.

[114] Dörfler, H., 1994: Grenzflächen- und Kolloidchemie, VCH Verlag, Weinheim.

[115] Schulman, S.G., 1985. Molecular Luminescence Spectroscopy – Methods and Applications: Part 1, John Wiley and Sons, Inc., New York.

[116] Binnig, G, Rohrer, H., Gerber, C., Weibel, E., 1982. Surface Studies by Scanning Tunneling Microscopy. Physical Review Letters 49, 57.

[117] Binnig, G., Rohrer, H., Gerber, C., Weibel, E., 1983. 7 x 7 reconstruction on Si(111) Resolved in Real Space. Physical Review Letters 50, 120.

[118] Pohl, D. W., Denk, W., Lanz, M., 1984. Optical stethoscopy: Image recording with resolution $\lambda/20$. Applied Physics Letters 44, 651.

[119] Lewis, A., Isaacson, M., Harootunian, A., Muray, M., 1984. Development of a 500 Å spatial resolution light microscope. Ultramicroscopy 13, 227.

[120] Binnig, G., Quate, C. F., Gerber, C., 1986. Atomic Force Microscopy. Physical Review Letters 56, 930.

[121] Binning, G., Quate, C.F., Gerber, C., 1986. Atomic Force Microscope. Physical Review Letters 56 (9), 930.

[122] Neubauer, G., Cohen, S.R., McClelland, G.M., Horne, D., Mate, C.M., 1990. Force Microscopy with a bidirectional capacitance sensor. Review of Scientific Instruments 61, 2296.

[123] McClelland, G.M., Erlandson, R., Chiang, S., 1987. in: Thompson, D. O., Chimenti, D. E. (Eds.) Review of Progress in Quantitative Non-Destructive Evaluation (6), 307.

[124] Itoh, Z., Suga, T. 1994. Piezoelectric Sensor for Detecting Force Gradients in Atomic Force Microscopy. Journal of Applied Physics 33, 334.

[125] Meyer, G., Amer, N.M., 1988. Novel optical approach to atomic force microscopy. Applied Physics Letters 53 (12), 1045.

[126] Meyer, G., Amer, N.M., 1988. Erratum: Novel optical approach to atomic force microscopy. Applied Physics Letters 53 (24), 2400.

[127] Alexander, S., Hellemans, L., Marti, O., Schneir, J., Elings, V., Hansma, P.K., Longmire, M., Gurley, J., 1989: An atomic-resolution atomic force microscope implemented using an optical lever. Journal of Applied Physied 65 (1), 164.

[128] Wolter, O., Bayer, Th., Geschner, J. 1991. Micromachined silicon sensors for scanning force microscopy. Journal of Vacuum Science and Technology B 9 (2), 1353.

[129] Albrecht, T.R., Akamine, S., Carver, T.E., Quate, C.F., 1990. Microfabrication of cantilever styli for the atomic force microscope. Journal of Vacuum Science and Technology A 8 (4), 3386.

[130] Cappella, B., Dietler, G., 1999. Force-distance curves by atomic force microscopy. Surface Science Reports 34, 1.

[131] Kikuwa, A., Hosaka, S., Honda, Y., Imura, R., 1995. Phase-locked noncontact scanning force microscope. Review of Scientific Instruments 66 (1), 101.

[132] T.R. Albrecht, P. Grütter, D. Horne, D. Rugar, 1991. Frequency modulation detection using high-Q cantilevers for enhanced force microscope sensitivity. Journal of Applied Physics 69 (2), 668.

[133] Zhong, Q., Innis, D., Kjoller, K., Elings, V.B., 1993. Fractured polymer/silica fiber surface studied by tapping mode atomic force microscopy. Surface Science 290, 668.

[134] Ho, H., West, P., 1996. Optimizing AC-Mode Atomic-Force Microscope Imaging. Scanning 18, 339.

[135] Cleveland, J.P., Anczykowski, B., Schmid, A.E., Elings, V.B., 1998. Energy dissipation in tapping-mode atomic force microscopy. Applied Physics Letters 72 (20), 2613.

[136] Claesson, P.M., Ederth, T., Bergeron, V., Rutland, M.W., 1996. Advances in Colloid and Interface Science 67, 1987.

[137] Ducker, W.A., Senden, T.J., Pashley, R.M., 1991. Direct measurement of colloidal forces using an atomic force microscope. Nature 353, 239.

[138] Butt, H.-J., 1991. Electrostatic Interaction in Atomic Force Microscopy. Biophysical Journal 60, 1438.

[139] Veeco probes website, http://www.veecoprobes.com

[140] Clifford, C.A., Seah, M.P., 2005. The determination of atomic force microscope cantilever spring constants via dimensional methods for nanomechanical analysis, Nanotechnology 16, 1666.

[141] Albrecht, T.R., Akaine, S., Carver, T.E., Quate, C.F., 1990. Microfabrication of cantilever styli for the atomic force microscope, Journal of Vacuum Science and Technology A8, 3386.

[142] Butt, H.-J., Siedle, P., Seifert, K., Fendler, K., Bember, E., Goldie, K., Engel, A., 1993. Scan speed limit in atomic force microscopy, Journal of Microscoscpy 169, 75.

[143] Sader, J.E., White, K., 1993. Theoretical analysis of the static deflection of plates for atomic force microscope applications, Journal of Applied Physics 74, 1.

[144] Sader, J. E., 1995. Parallel beam approximation for V-shaped atomic force cantilevers, Review of Scientific Instruments 66, 4583.

[145] Sader, J. E. 2002. Calibration of atomic force microscope cantilevers, in: A. Hubbard (Eds.), Encyclopedia of Surface and Colloid Science, Marcel Dekker, Inc., 846.

[146] Cok, S. M., Schaffer, T. E., Chynoweth, K. M., Wigton, M., Simmonds, R.W., Lang, K.M., 2006. Practical implementation of dynamic methods for measuring atomic force microscope cantilever spring constants. Nanotechnology 17, 2135.

[147] Burnham, N.A., Chen, X., Hodges, C.S., Matei, G.A., Thoreson, E.J., Roberts, C.J., Davies, M.C., Tendler, S.J.B., 2003. Comparsion of calibration methods for atomic-force microscopy cantilevers, Nanotechnology 14, 1.

[148] Cleveland, J. P., Manne, S., Bocek, D., Hansma, P.K., 1993. A nondestructive method for determining the spring constant of cantilevers for scanning force microscopy. Review of Scientific Instruments 64, 403.

[149] Sader, J. E., Chon, J.W.M., Mulvaney, P., White, L.R., 1999. Calibration of rectangular atomic force microscope cantilevers", Review of Scientific Instruments 70, 3967.

[150] Hutter, J. L., Bechhoefer, L., 1993. Calibration of atomic-force microscope tips. Review of Scientific Instruments 64, 1868.

[151] Torii, A., Sasaki, M., Hane, K., Okuma, S., 1996. A method for determining the spring constant of cantilevers for atomic force microscopy. Measurement Science and Technology 7, 179.

[152] Ohler,B. Practical Advice on the Determination of Cantilever Spring Constants.Veeco application note. Http:// www.veeco.com/pdfs/appnotes/AN94 Spring Constant Final_304.pdf

[153] Elimelech, M. 1989. Dissertation. Johns Hopkins University.

[154] Spielman, L. A., Friedlander, S. K., 1974. Role of Electrical Double Layer in Particle Deposition by Convective Diffusion. Journal of Colloid and Interface Science 46, 22.

[155] Seher, H., 2008. Personal communication.

[156] Rabung, T., 1998. Einfluß von Huminstoffen auf die Europium(III)-Sorption an Hämatit, Ph.D. Thesis, Universität Saarbrücken, Germany.

[157] Huber, F., 2007. Migration von Uranyl durch eine Quarz-Säule. Diplomarbeit am Institut für Nukleare Entsorgung, Forschungszentrum Karlsruhe, Deutschland.

[158] Schäfer, Th., Geckeis, H., Bouby, M., Fanghänel, Th., 2004. U, Th, Eu and colloid mobility in a granite fracture under near-natural flow conditions. Radiochmica Acta 92, 731.

[159] Flörsheimer, M., Kruse, K., Polly, R., Abdelmonem, A., Schimmelpfennig, B., Klenze, R., Fanghänel, Th., 2008. Hydration of Mineral Surfaces Probed at Molecular Level. Langmuir 24, 13434.

[160] Filby, A., Plaschke, M., Geckeis, H., Fanghänel, Th., 2008. Interaction of latex colloids with mineral surfaces and Grimsel granodiorite. Journal of Contaminant Hydrology 102, 273.

[161] Takahashi, Y., Kimura, T., Kato, Y., Minaia, Y., Tomingaa, T., 1997. Hydration structure of Eu(III) on aqueous ion-exchange resins using laser-induced fluorescence spectroscopy. Chemical Communications 2, 223.

[162] Ramirez-Aguilar, K.A., Rowlen, K.L., 1998. Tip Characterization from AFM Images of Nanometric Spherical Particles. Langmuir 14, 2562.

[163] Garcia, V.J., Martinez, L., Briceno-Valero, J.M., Schilling, C.H., 1997. Dimensional Metrology of Nanometric Spherical Particles using AFM. Probe Microscopy 1, 107.

[164] Kosmulski, M., 2001. Chemical properties of material surfaces. Marcel Dekker Inc., New York.

[165] Bergendahl, J., Grasso, D., 1999. Prediction of Colloid Detachment in a Model Porous Media: Thermodynamics. AIChE Journal, 45 (3), 475.

[166] Brow, C., Li, X., Ricka, J., Johnson, W.P., 2004. Comparison of microsphere deposition in porous media versus simple shear systems. Colloids and Surfaces A: Physicochemical and Engineering Aspects 253, 125.

[167] Tormoen, G. W., Drelich, J., J. 2005. Deformation of Soft Colloidal Probes during AFM Pull-off Force Measurements: Elimination of Nano-roughness Effects. Journal of Adhesion Science and Technology 19, 181.

[168] Considine, R. F., Hayes, R. A., Horn, R. G., 1999. Forces measured between latex spheres in aqueous electrolyte: Non-DLVO behaviour and sensitivity to dissolved gas Langmuir 15, 1657.

[169] Churchill, H., Teng, H., Hazen, R., 2004. Correlation of pH-dependent surface interaction forces to amino acid adsorption: Implications for the origin of life. American Mineralogist 89, 1048.

[170] Ryan, J.N., Elimelech, M., 1996. Colloid mobilization and transport in groundwater. Colloids and Surfaces A: Physicochemical and Engineering Aspects 107, 1.

[171] Brusseau, M.L., Wang, X., Hu, Q., 1994. Enhanced Transport of Low-Polarity Organic Compounds through Soil by Cyclodextrin. Environmental Science and Technology 28, 952.

[172] McDowell-Boyer, L.M., Hunt, J.R., Sitar, N., 1986. Particle Transport through Porous Media. Water Resources Research 22, 1901.

[173] Johnson, P.R., Sun, N., Elimelech, M., 1996. Colloid Transport in Geochemically Heterogeneous Porous Media: Modeling and Measurements. Environmental Science and Technology 30, 3284.

[174] Nazemifard, N., Masliyah, J.H., Bhattacharjee, S., 2006. Particle Deposition onto Charge Heterogeneous Surfaces: Convection – Diffusion – Migration Model. Langmuir 22, 9879.

[175] Duffadar, R.D., Davis, J.M., 2007. Interaction of micrometer-scale particles with nanotextured surfaces in shear flow. Journal of Colloid and Interface Science 308, 20.

[176] Song, L., Johnson, P.R., Elimelech, M., 1994. Kinetics of colloid deposition onto heterogeneously charged surfaces in porous media. Environmental Science and Technology 28, 1164.

[177] Adamczyk, Z., Dabros, T., Czarnecki, J., van de Ven, T.G.M., 1983. Particle transfer to solid surfaces. Advances in Colloid and Interface Science 19, 183.

[178] Hoek, E.M.V., Agarval, G.K., 2006. Extended DLVO interactions between spherical particles and rough surfaces. Journal of Colloid and Interface Science 298, 50.

[179] Kohn, M.J., Rakovan, J., Hughes, J.M., 2002. Phosphates: Geochemical, geobiological and materials importance. Mineralogical Society of America, Washington, D.C.

[180] Valsami-Jones, E., Ragnarsdottir, K.V., Putnis, A., Bosbach, D., Kemp, A.J., Cressey, G., 1998. The dissolution of apatite in the presence of aqueous metal cations at pH 2-7. Chemical Geology 151, 215.

[181] Rosso, Ribbe (Eds.) Phosphates- Geochemical, Geobiological and Materials importance. Reviews in Mineralogy and Geochemistry 48, 62.

[182] Schütz, W., Schubert, H., 1976. Der Einfluß von Anpreßkräften auf die Partikelhaftung. Chemie Ingenieur Technik 48/6, 567.

[183] Israelachvili, J.N., 1992. Intermolecular and Surface Forces, Academic Press, London.

[184] Bowen, W.R., Doneva, T.A., 2000. Atomic Force Microscopy Studies of Membranes: Effect of Surface Roughness on Double-Layer Interactions and Particle Adhesion. Journal of Colloid and Interface Science 229, 544.

[185] Assemi, S., Nalaskowski, J., Johnson, W. P., 2006. Direct force measurements between carboxylate-modified latex microspheres and glass using atomic force microscopy. Colloids and Surfaces A: Physicochemical and Engineering Aspects 286, 70.

[186] Pan, H. B., Darvell, B. W., 2007. Solubility of calcium fluoride and fluorapatite by solid titration. Archives of oral biology 52, 861.

[187] Greenland, D.J., 1971. Interaction between humic and fulvic acids and clays. Soil Science 111, 34.

[188] Plaschke, M., Römer, J., Klenze, R., Kim, J.I., 1999. In situ AFM study of sorbed humic acid colloids at different pH. Colloids and Surfaces A: Physicochemical and Engineering Aspects 160, 269.

[189] Wang, Y., Ferrari, M., 2000. Surface modification of micromachined silicon filters. Journal of Materials Science 35, 4923.

[190] Takahashi, Y., Kimura, T., Kato, Y., Minaia, Y., Tomingaa, T., 1997. Hydration structure of Eu(III) on aqueous ion-exchange resins using laser-induced fluorescence spectroscopy. Chemical Communications 2, 223.

[191] Blum, A. E., Lasaga, A.C., 1991. The role of surface speciation in the dissolution of albite. Geochimica et Cosmochimica Acta 55, 2193.

[192] Walther J., 1996. The relation between rates of alumino-silicate mineral dissolution, pH, temperature and surface charge. America Journal of Science 296, 693.

[193] Stumpf, S., Stumpf, Th., Lützenkirchen, J., Walter, C., Fanghänel, Th., 2008. Immobilization of trivalent actinides by sorption onto quartz and incorporation into siliceous bulk: Investigations by TRLFS. Journal of Colloid and Interface Science 318, 5.

[194] Arnold, Th., Zorn, T., Zänker, H., Bernhard, G., Nitsche, H., 2001. Sorption behaviour of U(VI) on phyllite: experiments and modelling. Journal of Contaminant Hydrology 47, 219.

[195] Ahmed, S. M., Van Cleave, A. B., 2009. Adsorption and flotation studies with quartz: Part I. Adsorption of calcium, hydrogen and hydroxyl ions on quartz. The Canadian Journal of Chemical Engineering 43 (1), 23.

[196] Fenter, P., Park, C., Sturchio, N. C., 2007. Adsorption of Rb^+ and Sr^{2+} at the orthoclase (001) interrface. Geochimica et Cosmochimica Acta 72 (7), 1848.

[197] Schlegel, M. L., Nagy, K. L., Fenter, P., Cheng, L., Sturchio, N. C., Jacobsen, S. D., 2006. Cation sorption on the muscovite (001) surface in chloride solutions using high-resolution x-ray reflectometry. Geochimica et Cosmochimica Acta 70 (14), 3549.

[198] Lyons, J.S., Furlong, D.N., Healy, T.W., 1981. The Electrical Double-Layer Properties of the Mica (Muscovite)-Aqueous Electrolyte Interface. Australian Journal of Chemistry 34, 1177.

[199] Debacher, N., Ottewill, R.H., 1991. An Electrokinetic examination of mica surfaces in aqueous media. Colloids and Surfaces 65, 51.

[200] Scales, P.J., Grieser, F., Healy, T.W., 1990. Electrokinetics of the muscovite mica aqueous solution interface. Langmuir 6, 582.

[201] Pashley, R.M., Israelachvili, J.N., 1984: Forces between mica surfaces in Mg^{2+}, Ca^{2+}, Sr^{2+}, and Ba^{2+} chloride solutions. Journal of Colloid and Interface Science 97, 446.

[202] Mahmood, T., Amirtharajah, A., Sturm, T. W., Dennett, K. E., 2001. A micromechanics approach for attachment and detachment of asymmetric colloidal particles. Colloids and Surfaces A: Physicochemical and Engineering Aspects 177, 99.

[203] Van den Hoven, Th. J. J., Bijsterbosch, B. H., 1987. Streaming Currents, Streaming Potentials and Conductances of Concentrated Dispersions of Negatively-Charged, Monodisperse Polystyrene Particles. Effect of Adsorbed Tetraalkylammonium Ions. Colloids and Surfaces 22, 187.

[204] Seebergh, J. E., Berg, J. C., 1995. Evidence of a hairy layer at the surface of polystyrene latex particles. Colloids and Surfaces A: Physicochemical and Engineering Aspects 100, 139.

[205] Ryan, J. N., Gschwend, P.M., 1993. Effects of Ionic strength and flow rate on colloid release: relating kinetics to intersurface potential energy. Journal of colloid and interface science 164, 21.

[206] Kallay, N., Biskup, B., Tomic, M., Matijevic, E., 1986. Particle adhesion and removal in model systems: X. The effect of electrolytes on particle detachment. Journal of Colloid and Interface Science 114, 357.

[207] Kallay, N., Barouch, E., Matijevic, E., 1987. Diffusional detachment of colloidal particles from solid/solution interfaces. Advances in Colloid and Interface Science 27, 1.

[208] Khilar, K. C., Fogler, H. S., 1984. The existence of a critical salt concentration for particle release. Journal of Colloid and Interface Science 100, 214.

[209] Kia, S. F., Fogler, H.S., Reed, M.G., 1987. Effect of pH on colloidally induced fines migration. Journal of Colloid and Interface Science 118, 158.

[210] Vaidya, R. N., Fogler, H.S., 1990. Formation damage due to colloidally induced fines migration. Colloids and Surfaces 50, 215.

[211] McDowell-Boyer, L.M., 1992. Chemical mobilization of micron-sized particles in saturated porous media under steady flow conditions. Environmental Science and Technology 26, 586.

[212] Cerda, C. M, 1987. Mobilization of kaolinite fines in porous media. Colloids and Surfaces 27, 219.

[213] Nasr-el-Din, H.A., Maini, B.B., Stanislav, P., 1991: Fines migration in unconsolidated sand formations.- AOSTRA Journal of Research, 7(1): 1.

[214] Khilar, K.C., Fogler, H.S., 1984. The existence of a critical salt concentration for particle release. Journal of Colloid and Interface Science 100, 214.

[215] Kolakowski, J.E., Matijevic, E., 1979. Particle adhesion and removal in model systems. I – Monodispersed chromium hydroxide on glass. Journal of the Chemical Society, Faraday Transactions Articles 75, 65.

[216] Kosakowsy, G., 2004. Anomalous transport of colloids and solutes in a shear zone. Journal of Contaminant Hydrology 72, 23.

[217] Vilks, P., Baik, M.-H., 2001. Laboratody migration experiments with radionuclides and natural colloids in a granite fracture. Journal of Contaminant Hydrology 47, 197.

[218] Degueldre,C., Grauer, R., Laube, A., Oess, A., Silby, A., 1996. Colloid properties in granitic groundwater systems. II: Stability and transport study. Applied Geochemistry 11, 697.

[219] Alonso, U., Missana, T., Patelli, A., Ceccato, D., Albarran, N., Garcia-Gutiérrez, Lopez-Torrubia, T., Rigato, V., 2009. Quantification of Au nanoparticles retention on a heterogeneous rock surface. Colloids and Surfaces A: Physicochemical and Engineering Aspects. In press.

[220] Seher, H., 2009. Personal communication.

[221] Huber, F., Seher, H., Kunze, P., Bouby, M., Banik, N. L., Hauser, W., Geckeis, H., Kienzler, B., Schäfer, Th., 2009. Laboratory study on colloid migration and colloid-radionuclide interaction under Grimsel groundwater conditions simulating glacial melt-water intrusion in the Äspö system (Part 1). Activity report "Colloid project". Forschungszentrum Karlsruhe, Germany.

[222] Missana, T., Alonso, U., Garcia-Guitérrez, M., Mingarro, M., 2008. Role of bentonite colloids on europium and plutonium migration in a granite fracture. Applied Geochemistry 23, 1484.

[223] Ryan, J.N., Elimelech, M., 1996. Colloid mobilization and transport in groundwater. Colloids and Surfaces A: Physicochemical and Engineering Aspects 107, 1.

VIII Appendix

1. List of figures

Fig. 1: Radiotoxicity of radionuclides in the spent fuel in dependence of time (see text)

Fig. 2: Size distributions of particles common in natural aquatic systems (see text)

Fig. 3: Schematic illustration of the adsorption of cations/anions onto a colloid surface as inner- or outer-sphere complexes (modified after Scheffer & Schachtschabel [51]) (see text)

Fig. 4: Scheme of colloid-mediated transport and colloidal interaction in subsurface environments (see text)

Fig. 5 (a) and (b): Schematic representation of the cross section of a metal oxide surface layer. Black spheres represent metal cations, white spheres are oxide ions [52] (see text)

Fig. 6: Layers of ions and potential distribution in the immediate vicinity of a negatively charged colloid particle dispersed in electrolyte solution [101] (see text)

Fig. 7: The Stern-Grahame model of the electrical double layer (see text)

Fig. 8: Electrostatic repulsion ΔG^{EL}, van der Waals attraction ΔG^{vdW} and Born repulsion ΔG^B energies and their total ΔG^T as a function of distance between similar charged particles dispersed in electrolyte at low (a) and high (b) electrolyte concentration [102] (see text)

Fig. 9: Formation of a contact angle between a liquid drop at the three phase liquid–solid–gas phase interface (see text)

Fig. 10: Light scattering in a colloid dispersion: due to the phase shifts the scattered light interferes and this leads to amplification or extinction of the scattered light (see text)

Fig. 11: Scheme of a fluorescence microscope (see text)

VIII Appendix 135

Fig. 12: Transmission vs. wavelength spectra of the exciter/emitter filters and the dichroic mirror (see text)

Fig. 13: Scheme of the SEM [92] (see text)

Fig. 14: Scheme of the AFM (see text)

Fig. 15: AFM colloid probe cantilever: the colloid particle (white sphere) is 1 µm in diameter (see text)

Fig. 16: Force-distance curve between a carboxylated latex sphere and a muscovite surface (pH 2, I = 10^{-2} M NaCl). The blue line represents approach of the cantilever, the red line retraction (see text)

Fig. 17: Force-distance curve between a carboxylated latex sphere and a muscovite surface (pH = 10 I = 10^{-2} M HCl). The blue line represents approach of the cantilever, the red line retraction (see text)

Fig. 18: Speciation of UO_2^{2+} in aqueous solution, in absence of CO_2. $[UO_2^{2+}] = 10^{-6}$ M, I = 10^{-2} NaCl [157] (see text)

Fig. 19: Zeta-potential measurements of fluorescent polystyrene colloids and bentonite (see text)

Fig. 20: Zeta-potentials of the carboxylated latex colloids as used in force spectroscopy and bentonite (see text)

Fig. 21: Zeta-potential of quartz in presence and absence of Eu(III) (see text)

Fig. 22: AFM image of muscovite surface with adsorbed latex colloids (pH = 4, I = 10^{-2} M NaCl, $c_{colloids}$ = 2 g/l, Eu(III) = 10^{-5} M) (see text)

VIII Appendix 136

Fig. 23: Fluorescence-optical image of muscovite (a) after adsorption of fluorescent colloids (pH = 4, I = 10^{-2} M NaCl, $c_{colloids}$ = 0.05 g/l) and (b) in the presence of Eu(III) (pH = 4, I = 10^{-2} M NaCl, $c_{colloids}$ = 0.05 g/l, Eu(III) = 10^{-5} M) (see text)

Fig. 24: Fluorescence intensity/surface coverage vs. pH for muscovite in the presence and absence of Eu(III): all measurements were undertaken at pH 2, 4, 6, 8, 10 (see text)

Fig. 25: Fluorescence intensity/surface coverage vs. pH of biotite in the presence and absence of Eu(III): all measurements were undertaken at pH 2, 4, 6, 8, 10 (see text)

Fig. 26: Fluorescence intensity/surface coverage vs. pH for albite in the presence and absence of Eu(III): all measurements were undertaken at pH 2, 4, 6, 8, 10 (see text)

Fig. 27: Fluorescence intensity/surface coverage vs. pH of K-feldspar in the presence and absence of Eu(III): all measurements were undertaken at pH 2, 4, 6, 8, 10 (see text)

Fig. 28: Fluorescence intensity/surface coverage vs. pH of quartz in the presence and absence of Eu(III): all measurements were undertaken at pH 2, 4, 6, 8, 10 (see text)

Fig. 29: Fluorescence intensity/surface coverage vs. pH of apatite in the presence and absence of Eu(III): all measurements were undertaken at pH 2, 4, 6, 8, 10 (see text)

Fig. 30: Fluorescence intensity/surface coverage vs. pH of sapphire in the presence and absence of Eu(III) (see text). A point diagram was chosen for a presentation of the results

Fig. 31: (a) SEM image and (b) fluorescence-optical image of the granodiorite surface (pH = 4, I = 10^{-2} M NaCl, $c_{colloids}$ = 0.05 g/l), both at approximately the same position (see text)

Fig. 32: (a) SEM-image and (b) fluorescence-optical image of the Grimsel granodiorite surface at approximately the same location. Sorption takes place preferably on the biotite edges, whereas fluorescence intensity on the planes is markedly weaker (pH = 4, I = 10^{-2} M NaCl, $c_{colloids}$ = 0.05 g/l, Eu(III) = 10^{-5} M) (see text)

VIII Appendix 137

Fig. 33: (a) Light-optical image and (b) fluorescence-optical image of a granodiorite surface site; fluorescence was detected on the entire apatite surface (pH 6, I = 10^{-2} M NaCl, $c_{colloids}$ = 0.05 g/l) (see text)

Fig. 34: Fluorescence intensity vs. time on the minerals albite and biotite (see text)

Fig. 35: Experimental snap-in forces on muscovite (see text)

Fig. 36: Experimental adhesion forces on muscovite (see text)

Fig. 37: Cantilever deflection vs. piezo position measured on muscovite with Grimsel groundwater (see text)

Fig. 38: Force-distance measurement (with baseline and hysteresis correction) measured on muscovite with Grimsel groundwater (see text)

Fig. 39: Force-distance approach curve measured on muscovite (see text)

Fig. 40: Experimental snap-in forces on biotite (see text)

Fig. 41: Experimental adhesion forces on biotite (see text)

Fig. 42: Experimental snap-in forces on K-feldspar (see text)

Fig. 43: Experimental adhesion forces on K-feldspar (see text)

Fig. 44: Force-distance approach curves measured on K-feldspar (see text)

Fig. 45: Experimental snap-in forces on quartz (see text)

Fig. 46: Experimental adhesion forces on quartz (see text)

Fig. 47: Force-distance approach curves measured on quartz (see text)

VIII Appendix 138

Fig. 48: Experimental snap-in forces on apatite (see text)

Fig. 49: Experimental adhesion forces on apatite (see text)

Fig. 50 (a) and (b): Height and deflection image of apatite after 10 min immersion in a solution of 10^{-2} M NaCl, pH 6 (see text)

Fig. 51 (a) and (b): Height and deflection image of apatite after 1 min immersion in a solution of 10^{-2} M NaCl, pH 2 (see text)

Fig. 52: Cantilever deflection vs. piezo position measured on titanite (Grimsel groundwater)

Fig. 53: Force-distance curve (with baseline and hysteresis correction) measured on titanite (Grimsel groundwater)

Fig. 54: Adhesion map measured on quartz. Values range from ca. 2.5 (dark areas) to 3.5 nN (bright areas) (see text)

Fig. 55: Comparison of experimental approach data (pH 3 and pH 6, I = 10^{-2} M NaCl) measured on quartz with DLVO calculations: the red line represents the DLVO fit to the measurement at pH 3, the black line represents the DLVO fit to the measurement at pH 6 (see text)

Fig. 56: Comparison of experimental approach data (pH = 6, 10^{-5} M Eu(III), I = 10^{-2} M NaCl and Grimsel groundwater) measured on quartz surface with DLVO calculations: the red line represents the DLVO fit to the measurement at pH 6, the black line represents the DLVO fit to the measurement with Grimsel groundwater (see text)

Fig. 57: Comparison of experimental approach data on muscovite surface with DLVO calculations (pH 2/6, I = 10^{-2} M NaCl): the red line represents the DLVO fit to the measurement at pH 2, the black line represents the DLVO fit to the measurement at pH 6 (see text)

VIII Appendix

Fig. 58: Comparison of experimental approach data on muscovite surface with DLVO calculations (pH 6, I = 10^{-2} M NaCl and Grimsel groundwater): the red line represents the DLVO fit to the measurement with Grimsel groundwater, the black line represents the DLVO fit to the measurement at pH 6 (see text)

Fig. 59: Comparison of experimental approach data on biotite surface with DLVO calculations (Grimsel groundwater) (see text)

Fig. 60: Comparison of experimental approach data on apatite surface with DLVO calculations (Grimsel groundwater) (see text)

2. List of tables

Table 1: Grimsel granodiorite bulk composition

Table 2: Surface tension components in [mN/m] of the liquids used for the contact angle measurements. Values were taken from van Oss [108]

Table 3: Specifications of cantilevers used in this work for imaging

Table 4: Specifications of cantilevers used in the force spectroscopy experiments

Table 5: Element concentration of Grimsel groundwater

Table 6: EDX data for the single minerals used in the sorption experiments (elemental composition in atom %, values < 1 % not included)

Table 7: Mineral surface roughness determined by AFM

Table 8: EDX data for the single minerals (elemental composition in atom %, values < 1 % not included) used in the force-spectroscopy experiments

Table 9: Adsorbing minerals and corresponding pH values on Grimsel granodiorite

VIII Appendix

Table 10: Results of the force-volume measurements; average values of 256 measurements per area

Table 11: Contact angles [°] measured on the different mineral surfaces with three independent liquids

Table 12: Calculated surface tension components and Hamaker constants of the mineral samples

Table 13: Calculated Hamaker constants for the mineral-water-colloid systems

Table 14: Fitted and measured surface potentials; values in brackets correspond to measured zeta- and streaming potential data

Table 15: Calculated maximum energy barriers for selected force spectroscopy measurements

VIII Appendix

3. List of symbols

Latin letters

Symbol	definition	unit
A	Area	m²
A_H	Hamaker constant	J
$A_{Hcolloid}$	Hamaker constant of the colloid	J
A_{HM}	Hamaker constant of the mineral	J
A_{Hwater}	Hamaker constant of water	J
A_S	signal amplitude	cps
b	distance from scattering centre	m
C	Coulomb	As
c	concentration	mol/L
C_{disp}	dispersion coefficient	Jm⁶
C_{ind}	induction coefficient	Jm⁶
c_M	mass concentration	g/L
C_{orient}	orientation coefficient	Jm⁶
D	diffusion coefficient	m²/s
d	particle diameter	m
d_{diff}	Debye-length	m
d_{op}	optical pathlength	m
E	field strength	V/m
-e	Electronic charge	C
F	force	N
F_{chem}	chemical force	N
F_{vdW}	van der Waals force	N
F_{El}	electrostatic double layer force	N

F_{cap}	capillary force	N
$f(\kappa R)$	Henry-function	-
G	free energy	J
h	Planck's constant	Js
I	ionic strength	mol/l
I_S	scattered light intensity	cps
I_0	intensity of incidental beam	cps
J	Joule	kg m²/s²
K	scattering vector	rad/m
k	spring constant	N/m
K_B	Boltzmann's constant	J/K
k_{ref}	spring constant (reference cantilever)	N/m
k_s	spring constant	N/m
M	concentration	mol/l
N	Newton	J/m
N_0	Avogadro's constant	1/mol
n	refraction index	-
P	Pressure	Pa
Q(0)	amplitude-weighed phase at time 0	rad
Q(t)	amplitude-weighed phase at time t	rad
q	auto correlation function	-
R	colloid radius	m
r	distance between atoms or molecules	m
r_{nom}	nominal colloid radius	m
r_p	hydrodynamic radius	m
S_{hard}	deflection sensitivity (hard surface)	m/V

S_{ref}	deflection sensitivity (reference cantilever)	m/V
T	Temperature	K
u_c	velocity of the collective movement	m/s
V	Volt	J/C
v_x	fluid velocity	m/s
W	interaction free energy	J
W_{disp}	dispersion interaction energy	J
W_{ind}	induction interaction energy	J
W_{orient}	orientation interaction energy	J
W_{vdW}	van der Waals interaction energy	J
x	distance between sphere and plate	m
x_d	cantilever deflection	m
z_i	valency if the ith ion sort	-

Greek letters

Symbol	definition	unit
α	Polarizability of atoms or molecules	C^2m^2/J
γ^{LW}	Lifshitz-van der Waals surface tension component	N/m
γ^+	surface tension electron acceptor component	N/m
γ^-	surface tension electron donor component	N/m
γ	Surface tension	N/m
γ_L	liquid surface tension	N/m
γ_S	surface energy of the solid	N/m
γ_{SL}	solid-liquid surface tension	N/m
ε	Permittivity of free space	C^2/Jm
ε_0	static dielectric constant of water	-
ε_x	extinction coefficient	-
ξ	zeta-potential	V
η	dynamic viscosity	g/ms

θ	contact angle	°
θ_a	scattering angle	°
κ	inverse Debye-length	m
κ_e	Electrical conductivity	A/Vm
λ	wavelength	m
λ_c	characteristic wavelength	m
μ_1, μ_2	dipole moment	Cm
μ_e	electrophoretic mobility	m²/Vs
ν	Frequency	1/s
ν_1, ν_2	ionization frequency of atoms/molecules	1/s
ν_p	particle velocity	m/s
ρ	number density of the atoms or molecules	/m³
ρ_d	density of particles	kg/m³
σ	standard deviation	
σ_c	collision diameter	m
τ	correlation time	s
Φ	intersurface potential energy	J
Φ_F	fluorescence quantum yield	cps
Ψ	surface potential	V
ω_v	dispersion frequency	1/s

4. Acronyms

Acronym	meaning
AFM	Atomic Force Microscope
CD-MUSIC	charge distribution mulit-site complexation
CEC	Cation Exchange Capacity
CRR	Colloid and radionuclide retardation experiment
DLVO	Derjaguin, Landau, Verwey, Overbeek
DMSO	dimethyl sulphoxide
DNA	desoxyribonucleic acid
EDX	Energy-dispersive x-ray fluorescence
GTS	Grimsel Test Site
ICP-MS	inductively coupled plasma mass spectrometry

VIII Appendix

IHP	inner Helmholtz plane
IUPAC	International Union of Pure and Applied Chemistry
LDA	Laser Doppler Anemometry
LSA	Linear Superposition Approximation
NOM	Natural Organic Matter
OHP	outer Helmholtz plane
PALS	phase analysis light scattering
PCS	Photon Correlation spectroscopy
pH_{pzc}	point of zero charge
PIXE	Particle-induced x-ray emission
RMS	round-mean-square
SEM	Scanning Electron Microscope
SPIP	Scanning Probe Image Processor
UV	ultraviolet radiation
XRD	x-ray diffraction

VDM Verlagsservicegesellschaft mbH

Die VDM Verlagsservicegesellschaft sucht für wissenschaftliche Verlage abgeschlossene und herausragende

Dissertationen, Habilitationen, Diplomarbeiten, Master Theses, Magisterarbeiten usw.

für die kostenlose Publikation als Fachbuch.

Sie verfügen über eine Arbeit, die hohen inhaltlichen und formalen Ansprüchen genügt, und haben Interesse an einer honorarvergüteten Publikation?

Dann senden Sie bitte erste Informationen über sich und Ihre Arbeit per Email an *info@vdm-vsg.de*.

Sie erhalten kurzfristig unser Feedback!

VDM Verlagsservicegesellschaft mbH
Dudweiler Landstr. 99   Telefon +49 681 3720 174
D - 66123 Saarbrücken   Fax   +49 681 3720 1749
www.vdm-vsg.de

Die VDM Verlagsservicegesellschaft mbH vertritt

Printed by Books on Demand GmbH, Norderstedt / Germany